LONDON MATHEMATICAL SOCIETY LECTURE NOTE SERIES

Editor: PROFESSOR G.C. SHEPHARD, University of East Anglia

This series publishes the records of lectures and seminars
on advanced topics in mathematics held at universities
throughout the world. For the most part, these are at post-
graduate level either presenting new material or describing
older matter in a new way. Exceptionally, topics at the
undergraduate level may be published if the treatment is
sufficiently original.

Prospective authors should contact the editor in the first
instance.

Already published in this series

continued overleaf

T0297313

London Mathematical Society Lecture Note Series. 29

Pontryagin Duality and the Structure of Locally Compact Abelian Groups

SIDNEY A. MORRIS

CAMBRIDGE UNIVERSITY PRESS

CAMBRIDGE

LONDON   NEW YORK   MELBOURNE

CAMBRIDGE UNIVERSITY PRESS
Cambridge, New York, Melbourne, Madrid, Cape Town, Singapore, São Paulo

Cambridge University Press
The Edinburgh Building, Cambridge CB2 8RU, UK

Published in the United States of America by Cambridge University Press, New York

www.cambridge.org
Information on this title: www.cambridge.org/9780521215435

First published 1977
Re-issued in this digitally printed version 2008

A catalogue record for this publication is available from the British Library

Library of Congress Cataloguing in Publication data

Morris, Sidney A   1947–
    Pontryagin duality and the structure of locally compact
abelian groups.

    (London Mathematical Society lecture note series; 29)
    Bibliography: p.
    Includes index.
    1. Locally compact Abelian groups. 2. Duality theory
(Mathematics) I. Title. II. Series; London Mathematical
Society. Lecture note series; 29.   QA387.M67   512'.55
76–53519

ISBN 978-0-521-21543-5 paperback

Contents

Preface

Most mathematicians are familiar with the fact that any
finitely generated abelian group can be expressed as a direct
product of cyclic groups. However, the equally attractive
generalization of this to topological groups is known only
to a small group of specialists. This is a real pity, as
the theory is not only elegant, but also a very pleasant
combination of topology and algebra. (It is also related to
some problems of diophantine approximation.)

Our aim is to describe the structure of locally compact
abelian groups and to acquaint the reader with the Pontryagin-
van Kampen duality theorem. This theorem is a deep result
and the standard proofs assume a knowledge of measure theory
and Banach algebras. In order to make the material accessible
to as large an audience as possible I make no such assumption.
Indeed even the amount of group theory and topology required
is small. Taking the Peter-Weyl theorem as known, I give a
simple and, as far as I know, new proof of the duality
theorem for compact groups and discrete groups. I then use
an approach analogous to that of D. W. Roeder (Category theory
applied to Pontryagin duality, *Pacific J.* 52 (1974) 519-527)
to extend the duality theorem to all locally compact abelian
groups. One of the beauties of this approach is that the
structure theory is derived simultaneously.

These notes are based on courses given in 1974 at the
University College of North Wales and in 1975 at the University
of New South Wales. The former was twenty lectures given to
an audience of graduate students and staff while the author
was a United Kingdom Science Research Council Senior Visiting

Fellow. The latter was twenty-eight lectures to a final honours class. In 1976 these notes were used as a reading course for final honours students at La Trobe University.

I am indebted to Ronald Brown for persuading me to give my first course on duality theory, for encouraging me to write the material for publication and for his helpful comments. I am extremely grateful to John Loxton and Rodney Nillsen for reading and criticizing the manuscript. Numerous other colleagues, in particular Peter Donovan and Peter Nickolas, made useful comments. I am also grateful to the students in the courses for removal or errors. I must record my thanks to Ian D. Macdonald for his moral support over a number of years and to Edwin Hewitt and Kenneth Ross for their willingness to answer by naive questions. I wish to thank Mesdames Rita Walker, Ulrike Bracken and Olwyn Bradford for their meticulous typing.

These notes are dedicated to Shnookie.

S.A.M.

La Trobe University
Melbourne
1976

1 · Introduction to topological groups

Definition. Let G be a set that is a group and a topological space. Then G is said to be a *topological group* if

(i) the mapping $(x,y) \to xy$ of $G \times G$ onto G is a continuous mapping of the cartesian product $G \times G$ (with the product topology) onto G ;

and (ii) the mapping $x \to x^{-1}$ of G onto G is continuous.

Examples:

(1) The additive group of real numbers with the "usual" topology (i.e. that given by the metric $d(x,y) = |x-y|$). It will be denoted by R .

(2) The multiplicative group of positive real numbers with the "usual" topology.

(3) The additive group of rational numbers with the "usual" topology – denoted by Q .

(4) The group of integers with the discrete topology (i.e. every set is an open set) – denoted by Z .

(5) Any group with the discrete topology.

(6) Any group with the indiscrete topology (i.e. the open sets are ϕ and the whole space).

(7) The "circle group" consisting of the complex numbers of modulus one (i.e. the set of numbers $e^{2\pi ix}$, $0 \leqslant x < 1$) with the group operation being multiplication of complex numbers and topology induced from that of the complex plane. This topological group is denoted by T (or S^1).

(8) *Linear groups*. Let $A = (a_{jk})$ be an $n \times n$ matrix, where the coefficients a_{jk} are complex numbers. The

1

transpose ${}^t A$ of the matrix A is the matrix (a_{kj})
and the conjugate $\bar{A}$ of A is the matrix $(\bar{a}_{jk})$,
where $\bar{a}_{jk}$ is the complex conjugate of the number
a_{jk} . The matrix A is said to be *orthogonal* if
$A = \bar{A}$ and ${}^t A = A^{-1}$ and *unitary* if $A^{-1} = {}^t(\bar{A})$.

The set of all non-singular $n \times n$ matrices (with complex number coefficients) is called the *general linear group*
(over the complex number field) and is denoted by $GL(n,C)$.
The subgroup of $GL(n,C)$ consisting of those matrices with
determinant one is the *special linear group* (over the complex field) and is denoted by $SL(n,C)$. The *unitary group*
$U(n)$ and the *orthogonal group* $O(n)$ consist of all unitary matrices and all orthogonal matrices, respectively;
they are subgroups of $GL(n,C)$. Finally we define the
special unitary group and the *special orthogonal group* as
$SU(n) = SL(n,C) \cap U(n)$ and $SO(n) = SL(n,C) \cap O(n)$,
respectively.

The group $GL(n,C)$ and all its subgroups can be regarded
as subsets of C^{n^2} (where C denotes the complex number
plane). As such $GL(n,C)$ and all its subgroups have induced topologies and it is easily verified that, with these,
they are topological groups.

Remark. Of course not every topology on a group makes it
into a topological group; i.e. the group structure and the
topological structure need not be compatible.

Example. Let G be the group of integers. Define a topology on G as follows: a subset U of G is open if

    (a) $0 \notin U$ or

    (b) $G \backslash U$ is finite.

Clearly this is a (compact Hausdorff) topology but Proposition 1 will show that G is not a topological group.

Proposition 1. *Let* G *be a topological group. For each* a ∈ G , *left and right translation by* a *are homeomorphisms of* G . *Inversion is also a homeomorphism.*

Proof. The map L_a: G → G given by g → ag is the product of the two continuous maps

   G → G × G given by g → (a,g) where a is fixed
and

   G × G → G given by (x,y) → xy
and is therefore continuous. So left translation by any a ∈ G is continuous. Further, L_a has a continuous inverse, namely $L_{a^{-1}}$, since $L_a[L_{a^{-1}}(g)] = L_a[a^{-1}g] = a(a^{-1}g) = g$ and $L_{a^{-1}}[L_a(g)] = L_{a^{-1}}[ag] = a^{-1}(ag) = g$. So left translation is a homeomorphism. Similarly right translation is a homeomorphism.

   The map I: G → G given by g → g^{-1} is continuous, by definition. Also I has a continuous inverse, namely I , itself, as $I[I(g)] = I[g^{-1}] = [g^{-1}]^{-1} = g$. So I is also a homeomorphism. //

   It is now clear that our example above is not a topological group as left translation by 1 takes the open set {-1} onto {0} , but {0} is not an open set. What we are really saying is that any topological group is a homogeneous space while the example is not.

Definition. A topological space X is said to be *homogeneous* if it has the property that for each ordered pair x,y of points of X , there exists a homeomorphism f: X → X such that f(x) = y .

   While every topological group is a homogeneous topological space, we will see shortly that not every homogeneous space can be made into a topological group.

Definition. A topological space is said to be a T_1-*space* if each point in the space is a closed set.

Definition. A topological space X is said to be *Hausdorff* or a T_2-*space* if for each pair of distinct points a and b in X , there exist open sets U_a and U_b , with $a \in U_a$, $b \in U_b$ and $U_a \cap U_b = \phi$.

It is readily seen that any Hausdorff space is a T_1-space but that the converse is false.

Example. Let X be any infinite set with the cofinite topology; that is, a subset U of X is open if and only if $U = X$, $U = \phi$ or $X \backslash U$ is finite.

Clearly this space is a T_1-space, but it is not Hausdorff as no (non-trivial) pair of open sets are disjoint.

We will see, however, that any topological group which is a T_1-space *is* Hausdorff. Incidentally, this is not true, in general, for homogeneous spaces - as the above example is homogeneous. As a consequence we will then have that not every homogeneous space can be made into a topological group.

Proposition 2. *Let* G *be any topological group and* e *its identity element. If* U *is any neighbourhood of* e , *then there exists an open neighbourhood* V *of* e *such that*

    (i) $V = V^{-1}$ (that is, V is *symmetric*)

    (ii) $V^2 \subseteq U$.

(Here $V^{-1} = \{v^{-1} : v \in V\}$ and

    $V^2 = \{v_1 v_2 : v_1 \in V$ and $v_2 \in V\}$ (*not* the set

                           $\{v^2 : v \in V\}$).)

Proof. Exercise.

(Use the continuity of $x \to x^{-1}$ at $x = e$, and the continuity of $(x,y) \to xy$ at $(x,y) = (e,e)$.) //

Proposition 3. *Any topological group* G *which is a* T_1-*space is also a Hausdorff space.*

Proof. Let x and y be distinct points of G . Then $x^{-1}y \neq e$. The set $G \setminus \{x^{-1}y\}$ is an open neighbourhood of e and so, by Proposition 2, there exists an open symmetric neighbourhood V of e such that $V^2 \subseteq G \setminus \{x^{-1}y\}$. Thus $x^{-1}y \notin V^2$.

Now xV and yV are open neighbourhoods of x and y , respectively. Suppose $xV \cap yV \neq \phi$. Then $xv_1 = yv_2$, where v_1 and v_2 are in V ; that is, $x^{-1}y = v_1 v_2^{-1} \in V.V^{-1} = V^2$ – which is a contradiction. Hence $xV \cap yV = \phi$ and so G is Hausdorff.//

So to check that a topological group is Hausdorff it is only necessary to verify that each point is a closed set. Indeed, by Proposition 1, it suffices to show that $\{e\}$ is a closed set.

Remark. Virtually all serious work on topological groups deals only with Hausdorff topological groups. (Indeed many authors include "Hausdorff" in their definition of topological group.) We will see one reason for this shortly. However, it is natural to ask: Does every group admit a Hausdorff topology which makes it into a topological group? The answer is obviously "yes" - the discrete topology. But we mention the following problem.

Question. Does every group admit a Hausdorff non-discrete group topology which makes it into a topological group?

S.Shelah (On a Kurosh problem: Jonsson groups; Frattini subgroups and untopologized groups) recently announced a negative answer, under the assumption of the continuum hypothesis. However in the special case that the group is abelian (= commutative) the answer is "yes" and to show this will be one of our earliest tasks.

EXERCISE SET ONE

1. Let G be a topological group, e its identity element, and k any element of G. If U is any neighbourhood of e, show that there exists an open neighbourhood V of e such that

(i) $V = V^{-1}$

and (ii) $V^2 \subseteq U$

and (iii) $k V k^{-1} \subseteq U$

(In fact, with more effort you can show that if K is a compact subset of G then V can be chosen to also have the property : (iv) for any $k \in K$, $k V k^{-1} \subseteq U$.)

2. (i)   Let G be any group and let $N = \{N\}$ be a family of normal subgroups of G. Show that the family of all sets of the form gN, as g runs through G and N runs through N is an open subbasis for a group topology on G. Such a topology is called a *subgroup topology*.

(ii)   Prove that *every* group topology on a finite group is a subgroup topology with N consisting of precisely one normal subgroup.

3. A topological space X is said to be a T_0-*space* if given any x and y in X, either there exists an open set containing x but not y, or there exists an open set containing y but not x. A topological space X is said to be *regular* if for each $x \in X$ and each open neighbourhood U of x, there exists a closed neighbourhood V of x such that $V \subseteq U$. Show that

(i)   any T_1-space is a T_0-space but that there exist T_0-spaces which are not T_1-spaces

(ii)   every topological group is a regular space

(iii)   any regular T_0-space is Hausdorff, and hence any

topological group which is a T_0-space is Hausdorff.

4. Let G be a topological group, A and B subsets
of G and g any element of G . Show that
(i)   If A is open then gA is open.
(ii)   If A is open and B is arbitrary, then AB is
       open.
(iii) If A and B are compact then AB is compact.
(iv)   If A is compact and B is closed then AB is
       closed.
(v)   If A and B are closed then AB need not be
       closed.

5. Let S be a compact subset of a metrizable topological group G , such that $xy \in S$ if x and y are in S .
Show that for each $x \in S$, $xS = S$. (Let y be a cluster
point of the sequence $x, x^2, x^3, \ldots$ in S and show that
$yS = \bigcap_{n=1}^{\infty} x^n S$; deduce that $yxS = yS$.) Hence show that S
is a subgroup of G . (Cf. Hewitt and Ross, *Abstract Har-
monic Analysis* I, Theorem 9.16.)

* * * * * * * * *

Definition.   Let G_1 and G_2 be topological groups. A
map $f\colon G_1 \to G_2$ is said to be a *continuous homomorphism* if
it is both a homomorphism of groups and continuous. If f
is also a homeomorphism then it is said to be a *topological
group isomorphism* or a *topological isomorphism* and G_1 and
G_2 are said to be *topologically isomorphic*.

Example.   Let R be the additive group of real numbers with
the usual topology and $R^\times$ the multiplicative group of positive real numbers with the usual topology. Then R and $R^\times$
are topologically isomorphic, where the topological isomor-

phism $R \to R^\times$ is $x \to \exp(x)$. (Hence we need not mention this group $R^\times$ again since, as topological groups, R and $R^\times$ are the same.)

Proposition 4. *Let* G *be a topological group and* H *a subgroup of* G. *With its relative topology as a subset of* G, H *is a topological group.*

Proof. The mapping $(x,y) \to xy$ of $H \times H$ onto H and the mapping $x \to x^{-1}$ of H onto H are continuous since they are restrictions of the corresponding mappings of $G \times G$ and G. //

Examples.
  (i) $Z \leqslant R$.
  (ii) $Q \leqslant R$.

Proposition 5. *Let* H *be a subgroup of a topological group* G. *Then*
  (i)   *the closure* $\bar{H}$ *of* H *is a subgroup of* G ;
  (ii) *if* H *is a normal subgroup of* G *then* $\bar{H}$ *is a normal subgroup of* G ;
  (iii) *if* G *is Hausdorff and* H *is abelian, then* $\bar{H}$ *is abelian.*

Proof. Exercise.

Corollary. *Let* G *be a topological group. Then* $\overline{\{e\}}$ *is a closed normal subgroup of* G ; *indeed, it is the smallest closed subgroup of* G. *If* $g \in G$, *then* $\overline{\{g\}}$ *is the coset* $g\overline{\{e\}} = \overline{\{e\}}g$. *(Of course if* G *is Hausdorff then* $\overline{\{e\}} = \{e\}$ *.)*

Proof. This follows immediately from Proposition 5 (ii) by

8

noting that {e} is a normal subgroup of G . //

Proposition 6. *Any open subgroup H of a topological group G is (also) closed.*

Proof. Let x_i , $i \in I$ be a set of right coset representatives of H in G . So $G = \bigcup_{i \in I} Hx_i$, where $Hx_i \cap Hx_j = \phi$, for any distinct i and j in the index set I . Since H is open, so is Hx_i open, for each $i \in I$. Of course for some $i_0 \in I$, $Hx_{i_0} = H$, that is, we have

$$G = H \cup \left[\bigcup_{i \in J} Hx_i \right] , \text{ where } J = I \setminus \{i_0\} .$$

These two terms are disjoint and the second term, being the union of open sets, is open. So H is the complement (in G) of an open set, and is therefore closed in G . //

Note that the converse of Proposition 6 is false. For example, Z is a closed subgroup of R , but it is not an open subgroup of R .

Proposition 7. *Let H be a subgroup of a Hausdorff group G . If H is locally compact, then H is closed in G . In particular this is the case if H is discrete.*

Proof. Let K be a compact neighbourhood in H of e . Then there exists a neighbourhood U in G of e such that $U \cap H = K$. In particular, $U \cap H$ is closed in G . Let V be a neighbourhood in G of e such that $V^2 \subseteq U$.

If $x \in \bar{H}$, then as $\bar{H}$ is a group (Proposition 5), $x^{-1} \in \bar{H}$. So there exists an element $y \in Vx^{-1} \cap H$. We will show that $yx \in H$. As $y \in H$, this will imply that $x \in H$ and hence H is closed, as required.

To show that $yx \in H$ we verify that yx is a limit point of $U \cap H$. As $U \cap H$ is closed this will imply that $yx \in U \cap H$ and so, in particular, $yx \in H$.

Let 0 be an arbitrary neighbourhood of yx . Then $y^{-1}0$ is a neighbourhood of x , and so $y^{-1}0 \cap xV$ is a neighbourhood of x . As $x \in \bar{H}$, there is an element $h \in (y^{-1}0 \cap xV) \cap H$. So $yh \in 0$. Also $yh \in (Vx^{-1})(xV) = V^2 \subseteq U$, and $yh \in H$; that is, $yh \in 0 \cap (U \cap H)$. As 0 is arbitrary, this says that yx is a limit point of $U \cap H$, as required. //

Proposition 8. *Let* U *be a symmetric neighbourhood of* e *in a topological group* G . *Then* $H = \overset{\infty}{\underset{n=1}{U}} U^n$ *is an open (and closed) subgroup of* G .

Proof. Clearly H is a subgroup of G . Let $h \in H$. Then $h \in U^n$, for some n . So $h \in hU \subseteq U^{n+1} \subseteq H$; that is, H contains the neighbourhood hU of h . As h was an arbitrary element of H , H is open in G . It is also closed in G , by Proposition 6. //

Corollary 1. *Let* U *be any neighbourhood of* e *in a connected topological group* G . *Then* $G = \overset{\infty}{\underset{n=1}{U}} U^n$; *that is, any connected group is generated by any neighbourhood of* e .

Proof. Let V be a symmetric neighbourhood of e such that $V \subseteq U$. By Proposition 8, $H = \overset{\infty}{\underset{n=1}{U}} V^n$ is an open and closed subgroup of G .

As G is connected, $H = G$; that is $G = \overset{\infty}{\underset{n=1}{U}} V^n$. As $V \subseteq U$, $V^n \subseteq U^n$, for each n and so $G = \overset{\infty}{\underset{n=1}{U}} U^n$, as required. //

Definition. A topological group G is said to be *compactly generated* if there exists a compact subset X of G such that G is the smallest subgroup (of G) containing X .

Examples:

(i)   R is compactly generated by $[0,1]$ (or any other non-trivial compact interval).

(ii)   Of course, any compact group is compactly generated.

Corollary 2. *Any connected locally compact group is compactly generated.*

Proof.   Let K be any compact neighbourhood of e . Then by Corollary 1, $G = \bigcup_{n=1}^{\infty} K^n$; that is, G is compactly generated. //

Remark. An objective of this course of lectures is to describe the structure of compactly generated locally compact Hausdorff abelian groups. We now see that this class includes all connected locally compact Hausdorff abelian groups.

Notation: *LCA-group* $\equiv$ locally compact Hausdorff abelian topological group.

Proposition 9.   *The component of the identity (that is, the largest connected subset containing e) of a topological group is a closed normal subgroup.*

Proof.   Let C be the component of the identity in a topological group G . As in any topological space components are closed sets, C is closed. Let $a \in C$. Then $a^{-1}C \subseteq C$ as $a^{-1}C$ is connected (being a homeomorphic image of C) and contains e . So $\bigcup_{a \in C} a^{-1}C = C^{-1}C \subseteq C$ - which implies that C is a subgroup. To see that C is a normal subgroup, simply note that for each x in G , $x^{-1}Cx$ is a connected set containing e and so $x^{-1}Cx \subseteq C$. //

Proposition 10. *Let* N *be a normal subgroup of a topo-*
logical group G . *If the quotient group* G/N *is given*
the quotient topology under the canonical homomorphism
p: G $\rightarrow$ G/N *(that is,* U *is open in* G/N *if and only if*
$p^{-1}(U)$ *is open in* G *), then* G/N *becomes a topological*
group. Further, the map p *is not only continuous but*
also open. (A map is said to be *open* if the image of
every open set is open.)

Proof. The verification that G/N with the quotient
topology is a topological group is routine. That the map
p is continuous is obvious (and true for all quotient maps
of topological spaces).

To see that p is an open map, let O be an open set
in G . Then $p^{-1}(p(O))$ = NO $\subseteq$ G . Since O is open, NO
is open. (See Exercise Set One, Problem 4.) So by the
definition of the quotient topology on G/N , p(O) is open
in G/N ; that is, p is an open map. //

Remarks.
(i)   Note that quotient maps of topological spaces are not
      necessarily open maps.
(ii) Quotient maps of topological groups are not necessarily
      closed maps. For example, if R^2 denotes the product
      group R × R with the usual topology, and p is the
      projection of R^2 onto its first factor R , then the
      set S = {(x, $\frac{1}{x}$): x $\in$ R , x $\neq$ 0} is closed in R^2
      and p is a quotient map with p(S) not closed in
      R .

Proposition 11. *If* G *is a topological group and* N *is*
a compact normal subgroup of G *then the canonical homo-*
morphism p: G $\rightarrow$ G/N *is a closed map. The homomorphism* p
is also an open map.

Proof. If S is a closed subset of G , then $p^{-1}(p(S))$ = NS - the product in G of a compact set and a closed set, which by Exercise Set One, Problem 4, is a closed set. So p(S) is closed in G/N and p is a closed map. As p is a quotient mapping, Proposition 10 implies that it is an open map. //

Definition. A topological space is said to be *totally disconnected* if the component of each point is the point itself.

Proposition 12. *If G is any topological group and C is the component of the identity, then G/C is a totally disconnected topological group.*

Proof. Note that C is a normal subgroup of G and so G/C is a topological group.

The proof that G/C is totally disconnected is left as an exercise.

Proposition 13. *If G/N is any quotient group of a locally compact group G , then G/N is locally compact.*

Proof. Simply observe that any open continuous image of a locally compact space is locally compact. //

Proposition 14. *Let G be a topological group and N a normal subgroup. Then G/N is discrete if and only if N is open. Also G/N is Hausdorff if and only if N is closed.*

Proof. Obvious (noting that a T_1-group is Hausdorff). //

EXERCISE SET TWO

1. Let G and H be topological groups and f: G → H a homomorphism. Show that f is continuous if and only if it is continuous at the identity; that is, if and only if for each neighbourhood U in H of e , there exists a neighbourhood V in G of e such that $f(V) \subseteq U$.

2. Show that the circle group T is topologically isomorphic to the quotient group R/Z .

3. Let B_1 and B_2 be (real) Banach spaces. Verify that

(i) B_1 and B_2 , with the topologies determined by their norms, are topological groups.

(ii)   If T: B_1 → B_2 is a continuous homomorphism (of topological groups) then T is a continuous linear transformation. (So if B_1 and B_2 are "isomorphic as topological groups" then they are "isomorphic as topological vector spaces" but not necessarily "isomorphic as Banach spaces".)

4. Let H be a subgroup of topological group G . Show that H is open in G if and only if H has non-empty interior (that is, if and only if H contains an open subset of G).

5. Let H be a subgroup of a topological group G . Show that

(i) $\bar{H}$ is a subgroup of G .

(ii)   If H is a normal subgroup of G , then $\bar{H}$ is a normal subgroup of G .

(iii)   If G is Hausdorff and H is abelian, then $\bar{H}$ is abelian.

6. Let Y be a dense subspace of a Hausdorff space X .
If Y is locally compact show that Y is open in X .
Hence show that a locally compact subgroup of a Hausdorff
group is closed.

7. Let C be the component of the identity in a topological group G . Show that G/C is a Hausdorff totally
disconnected topological group. Further show that if f is
any continuous homomorphism of G into any totally disconnected topological group H , then there exists a continuous homomorphism g: G/C → H such that gp = f , where
p is the projection p: G → G/C .

8. Show that the commutator subgroup C of a connected
topological group G is connected. (C is generated by
$\{g_1^{-1}g_2^{-1}g_1g_2: g_1,g_2 \in G\}$.)

9. If H is a totally disconnected normal subgroup of
a connected Hausdorff group G , show that H lies in the
centre of G (that is, gh = hg , for all g ∈ G and
h ∈ H).
(Hint: Fix h ∈ H and observe that the map $g → ghg^{-1}$ takes
G into H .)

10. (i)   Let G be any topological group. Verify that
          G/cl{e} is a Hausdorff topological group,
          where cl{e} denotes the closure in G of
          the identity. Show that if H is any Haus-
          dorff group and f: G → H is a continuous
          homomorphism, then there exists a continuous
          homomorphism g: G/cl{e} → H such that
          gp = f , where p is the canonical map
          p: G → G/cl{e} .
          (This result is the usual reason given for

studying Hausdorff topological groups rather
than arbitrary topological groups. However,
the following result which says in effect
that all of the topology of a topological
group lies in its "Hausdorffization" (namely
$G/cl\{e\}$) is perhaps a better reason.)

(ii) Let G_i denote the group G with the indiscrete topology and $i: G \to G_i$ the identity
map. Verify that the map $p \times i: G \to G/cl\{e\}$
$\times G_i$, given by $p \times i(g) = (p(g),i(g))$, is
a topological group isomorphism of G onto
its image $p \times i(G)$.

11. Show that every Hausdorff group H is topologically
isomorphic to a closed subgroup of an arcwise connected,
locally arcwise connected Hausdorff group G. (Consider
the set G of all functions $f: [0,1) \to H$ such that there
is a sequence $0 = a_0 < a_1 < a_2 < \dots < a_n = 1$ with f
being constant on each $[a_k,a_{k-1})$. Define a group structure
on G by $fg(t) = f(t)g(t)$ and $f^{-1}(t) = (f(t))^{-1}$, where
f and $g \in G$ and $t \in [0,1)$. The identity of G is the
function identically equal to e in G. For $\varepsilon > 0$ and
any neighbourhood V of e in G let $U(V,\varepsilon)$ be the set
of all f such that $\lambda(\{t \in [0,1): f(t) \notin V\}) < \varepsilon$, where
λ is Lebesgue measure on $[0,1)$. The set of all $U(V,\varepsilon)$
is an open basis for a group topology on G. The constant
functions form a closed subgroup of G topologically isomorphic to H.)

* * * * * * * * *

Remarks on Products. Let $\{G_i: i \in I\}$ be a family of
sets, for some index set I. The *direct product (cartesian
product)* of the family $\{G_i: i \in I\}$ is the set which con-

sists of elements $\underset{i\in I}{\Pi}\ g_i$, with each $g_i \in G_i$, and is
denoted by $\underset{i\in I}{\Pi}\ G_i$. If each G_i is a topological space
then the *product topology (Tychonoff topology)* is that
topology which has as a basis for its open sets the
collection of all $\underset{i\in I}{\Pi}\ U_i$, where each U_i is open in
G_i , and all but a *finite* number of $U_i = G_i$. (Note that
the product topology is quite different from the topology
which at first sight seems more natural – the *box topology*,
which has as a basis for its open sets the collection of all
$\underset{i\in I}{\Pi}\ U_i$, where U_i is open in G_i .) Observe that the
product topology is the coarsest topology on $\underset{i\in I}{\Pi}\ G_i$ for
which each of the canonical projections $p_i \colon \underset{i\in I}{\Pi}\ G_i \to G_i$ is
continuous. The most important result on product topologies
is the following:

Tychonoff Theorem. *If* $\{G_i \colon i \in I\}$ *is a family of compact*
topological spaces then $\underset{i\in I}{\Pi}\ G_i$ *, with the product topology,*
is compact.

Since each $p_i \colon \underset{i\in I}{\Pi}\ G_i \to G_i$ is continuous, if $\underset{i\in I}{\Pi}\ G_i$
(with the product topology) is compact then each G_i is
compact.

Note that the Tychonoff theorem would be false if
"product topology" were replaced by "box topology". (For
example, if each G_i is a finite discrete space then
$\underset{i\in I}{\Pi}\ G_i$ with the box topology is discrete – hence not
compact, if I is infinite!)

If each G_i is a group then $\underset{i\in I}{\Pi}\ G_i$ has the obvious
group structure ($\underset{i\in I}{\Pi}\ g_i \cdot \underset{i\in I}{\Pi}\ h_i = \underset{i\in I}{\Pi}\ (g_i h_i)$, where g_i
and $h_i \in G_i$).

If $\{G_i \colon i \in I\}$ is a family of groups then the
restricted direct product (weak direct product), denoted
$\underset{i\in I}{\Pi^r}\ G_i$, is the subgroup of $\underset{i\in I}{\Pi}\ G_i$ consisting of elements
$\underset{i\in I}{\Pi}\ g_i$, with $g_i = e$, for all but a finite number of
$i \in I$.

From now on, if $\{G_i : i \in I\}$ is a family of topological groups then $\underset{i \in I}{\Pi} G_i$ will denote the direct product *with the product topology*. Further $\underset{i \in I}{\Pi^r} G_i$ will denote the restricted direct product with the topology induced as a subspace of $\underset{i \in I}{\Pi} G_i$.

Proposition 15.   *If each G_i , $i \in I$ is a topological group then $\underset{i \in I}{\Pi} G_i$ is a topological group. Further $\underset{i \in I}{\Pi^r} G_i$ is a dense subgroup of $\underset{i \in I}{\Pi} G_i$.*

Proposition 16.   *Let $\{G_i : i \in I\}$ be a family of topological groups. Then*

(i) $\underset{i \in I}{\Pi} G_i$ *is locally compact if and only if each G_i is locally compact and all but a finite number of G_i are compact.*

(ii) $\underset{i \in I}{\Pi^r} G_i$ *is locally compact Hausdorff if and only if each G_i is locally compact Hausdorff and $G_i = \{e\}$ for all but a finite number of G_i .*

Proof.   Exercise.

To prove the result we promised - every infinite abelian group admits a non-discrete Hausdorff group topology - we need some basic group theory.

Definition.   A group D is said to be *divisible* if for $n = 1, 2, \ldots,$ $\{x^n : x \in D\} = D$; that is, every element of D has an nth root.

Examples: R , T but not Z .

Proposition 17.   *Let H be a subgroup of an abelian group G . If ϕ is any homomorphism of H into a divisible abelian group D , then ϕ can be extended to a homomorphism of G into D .*

18

Proof. By Zorn's lemma it suffices to show that if $x \notin H$, ϕ can be extended to the group $H_0 = \{x^n h: h \in H , n \in Z \}$.

Case (i). $x^n \notin H$, $n = 1,2,\ldots$ Then define $\phi(x^n h) = \phi(h)$. Clearly ϕ is well-defined, a homomorphism, and extends ϕ on H .

Case (ii). Let $k \geqslant 2$ be the *least* positive integer n such that $x^n \in H$. So $\phi(x^k) = d \in D$. As D is divisible, there is a $z \in D$ such that $z^k = d$. Define $\phi(x^n h) = \phi(h) z^n$. Clearly ϕ is well-defined, a homomorphism and extends ϕ on H . //

Corollary. *If G is an abelian group then for any g and h in G , with $g \neq h$, there exists a homomorphism $\phi: G \to T$ such that $\phi(g) \neq \phi(h)$; that is, ϕ separates points of G .*

Proof. Clearly it suffices to show that for each $g \neq e$ in G there exists a homomorphism $\phi: G \to T$ such that $\phi(g) \neq e$.

Case (i). $g^n = e$, and $g^k \neq e$ for $0 < k < n$. Let $H = \{g^m: m = 0,\pm1,\pm2,\ldots\}$. Define $\phi: H \to T$ by $\phi(g) =$ an nth root of unity $= r$, say, $(r \neq e)$, and $\phi(g^m) = r^m$, for each m . Now extend ϕ to G by Proposition 17.

Case (ii). $g^n \neq e$, for any n . Define $\phi(g) = z$, for any $z \neq e$ in T . Extend ϕ to H and then, by Proposition 17, to G . //

For later use we also record the following corollary of Proposition 17.

Proposition 18. *Let H be an open divisible subgroup of an abelian topological group G . Then G is topologically isomorphic to $H \times G/H$. (Of course, G/H is a discrete topological group.)*

Proof. Exercise.

Theorem 1. *If* G *is any infinite abelian group, then* G *admits a non-discrete Hausdorff group topology.*

Proof. Let $\{\phi_i : i \in I\}$ be the family of distinct homomorphisms of G into T . Put $H = \prod_{i \in I} T_i$, where each $T_i = T$. Define a map $f : G \to H = \prod_{i \in I} T_i$ by putting $f(g) = \prod_{i \in I} \phi_i(g)$. Since each ϕ_i is a homomorphism, f is also a homomorphism. By the Corollary of Proposition 17, f is also one-one; that is, G is isomorphic to the subgroup f(G) of H .

As H is a Hausdorff topological group, f(G) , with the topology induced from H , is also a Hausdorff topological group. It only remains to show that f(G) is not discrete.

If f(G) were discrete then, by Proposition 7 it would be a closed subgroup of H . But by the Tychonoff theorem H is compact and so f(G) would be compact; that is, f(G) would be an infinite discrete compact space – which is impossible. So f(G) is not discrete. //

Remark. The Corollary of Proposition 17 was essential to the proof of Theorem 1. The Corollary is a special case of the following theorem, which shall be discussed in Chapter 6.

Theorem. *If* G *is any LCA-group, then for any* g *and* h *in* G , *with* $g \neq h$, *there exists a continuous homomorphism* $\phi : G \to T$ *such that* $\phi(g) \neq \phi(h)$.

EXERCISE SET THREE

1. If $\{G_i : i \in I\}$ is a family of topological groups, show that

(i) $\prod_{i \in I} G_i$ is a topological group.

(ii) $\prod_{i \in I}^r G_i$ is a dense subgroup of $\prod_{i \in I} G_i$.

(iii) $\prod_{i \in I} G_i$ is locally compact if and only if each G_i
is locally compact and all but a finite number of G_i
are compact.

(iv) $\prod_{i \in I}^{r} G_i$ is locally compact Hausdorff if and only if
each G_i is locally compact Hausdorff and $G_i = \{e\}$
for all but a finite number of G_i .

2. Show that if G is an abelian topological group with
an open divisible subgroup H , then G is topologically
isomorphic to H × G/H .

3. Let G be a torsion-free abelian group (that is,
$g^n \neq e$ for each $g \neq e$ in G , and each positive integer
n). Show that if g and h are in G with $g \neq h$, then
there exists a homomorphism ϕ of G into R such that
$\phi(g) \neq \phi(h)$.

4. Let G be a locally compact totally disconnected
topological group.

(i)   Show that there is a neighbourhood base of the
        identity consisting of compact open subgroups. (You
        may assume that any locally compact Hausdorff totally
        disconnected topological space has a base for its
        topology consisting of compact open sets.)

(ii)   If G is compact, show that the "subgroups" in (i)
        can be chosen to be normal.

(iii)   Hence show that any compact totally disconnected
        topological group is topologically isomorphic to a
        closed subgroup of a product of finite discrete groups.
        (Let $\{A_i : i \in I\}$ be a base of neighbourhoods of the
        identity consisting of open normal subgroups. Let
        $\phi_i : G \to G/A_i$, $i \in I$, be the canonical homomorphisms,
        and define $\Phi: G \to \prod_{i \in I} (G/A_i)$ by putting
        $\Phi(g) = \prod_{i \in I} \phi_i(g_i)$.)

5. Let $f: R \to T$ be the canonical map and θ any irrational number. On the topological space $G = R^2 \times T^2$ we define an operation

$$(x_1, x_2, t_1, t_2) \cdot (x_1', x_2', t_1', t_2')$$
$$= (x_1 + x_1', x_2 + x_2', t_1 + t_1' + f(x_2 x_1'), t_2 + t_2' + f(\theta x_2 x_1')) .$$

Show that, with this operation, G is a topological group and that the commutator subgroup of G is not closed in G.

6. Let I be a set directed by a partial ordering $\leq$. For each $i \in I$, let there be given a Hausdorff topological group G_i. Suppose that for each i and j in I such that $i < j$, there is an open continuous homomorphism f_{ji} of G_j into G_i. Suppose further that if $i < j < k$ then $f_{ki} = f_{ji} f_{kj}$. The object consisting of I, the groups G_i and the mappings f_{ji}, is called an *inverse mapping system*. The subset H of the product group $G = \underset{i \in I}{\Pi} G_i$ consisting of all (x_i) such that if $i < j$ then $x_i = f_{ji}(x_j)$ is called the *projective limit* of the inverse mapping system. Show that H is a closed subgroup of G.

* * * * * * * * *

Theorem 2 (Baire Category Theorem). *If* X *is a locally compact regular space then* X *is not the union of a countable collection of closed sets all having empty interior.*

Proof. Suppose that $X = \overset{\infty}{\underset{n=1}{U}} A_n$, where each A_n is closed and $\text{Int.}(A_n) = \phi$, for each n. Put $D_n = X \backslash A_n$. Then each D_n is open and dense in X. We will show that $\overset{\infty}{\underset{n=1}{\cap}} D_n \neq \phi$, contradicting the equality $X = \overset{\infty}{\underset{n=1}{U}} A_n$.

Let U_0 be a non-empty open subset of X such that $\bar{U}_0$ is compact. As D_1 is dense in X, $U_0 \cap D_1$ is a non-empty open subset of X. Using the regularity of X we can choose a non-empty open set U_1 such that $\bar{U}_1 \subseteq U_0 \cap D_1$. Inductively define U_n so that each U_n is a non-empty open set and $\bar{U}_n \subseteq U_{n-1} \cap D_n$. Since U_0 is compact and each $\bar{U}_n$ is non-empty, $\bigcap_{n=1}^{\infty} \bar{U}_n \neq \phi$. This implies $\bigcap_{n=1}^{\infty} D_n \neq \phi$. //

Remark. The above Theorem remains valid if "locally compact regular" is replaced by "complete metric" or "locally compact Hausdorff".

Corollary. *Let* G *be any countable locally compact Hausdorff topological group. Then* G *has the discrete topology.*

Proof. Exercise.

Theorem 3 (Open Mapping Theorem). *Let* G *be a locally compact group which is σ-compact; that is,* $G = \bigcup_{n=1}^{\infty} A_n$, *where each* A_n *is compact. Let* f *be any continuous homomorphism of* G *onto a locally compact Hausdorff group* H . *Then* f *is an open mapping.*

Proof. Let U be the family of all symmetric neighbourhoods of e in G and U' the family of all neighbourhoods of e in H . It suffices to show that for every $U \in U$ there is a $U' \in U'$ such that $U' \subseteq f(U)$.

Let $U \in U$. Then there exists a $V \in U$ having the property that $\bar{V}$ is compact and $(\bar{V})^{-1}\bar{V} \subseteq U$. The family of sets $\{xV : x \in G\}$ is then an open cover of G and hence also of each compact set A_n . So a finite collection of these sets will cover any given A_n . Thus there is a countable collection $\{x_n V : n = 1, 2, \ldots\}$ which covers G .

So $H = \bigcup_{n=1}^{\infty} f(x_n V) = \bigcup_{n=1}^{\infty} f(x_n \bar{V}) = \bigcup_{n=1}^{\infty} f(x_n) f(\bar{V})$. This expresses H as a countable union of closed sets, and by the Baire Category Theorem, one of them must have non-empty interior; that is, $f(x_m) f(\bar{V})$ contains an open set. Then $f(\bar{V})$ contains an open subset V' of H .

To complete the proof select any point x' of V' and put $U' = (x')^{-1} V'$. Then we have

$$U' = (x')^{-1} V' \subseteq (V')^{-1} V' \subseteq (f(\bar{V}))^{-1} f(\bar{V}) = f((\bar{V})^{-1} \bar{V}) \subseteq f(U) ,$$

as required. //

EXERCISE SET FOUR

1. Prove the Baire Category Theorem for complete metric spaces.

2. Show that any countable locally compact Hausdorff group has the discrete topology.

3. Show that the Open Mapping Theorem does not remain valid if either of the conditions "σ-compact" or "onto" is deleted.

4. Show that any continuous homomorphism of a compact group onto a Hausdorff group is open.

5. Show that for any positive integer n , T^n is topologically isomorphic to R^n/Z^n .

6. (i) Let ϕ be a homomorphism of a topological group G into a topological group H . If X is a non-empty subset of G such that the restriction $\phi: X \to H$ is an open map, show that $\phi: G \to H$ is also an open map.
   (Hint: For any subset U of G ,
   $$\phi(U) = \bigcup_{g \in G} \phi(U \cap gX) \ .)$$
   (ii) Hence show that if G and H are locally compact Hausdorff groups with ϕ a continuous

homomorphism $\phi : G \to H$ such that for some
compact subset K of G, $\phi(K)$ generates
H algebraically, then ϕ is an open map.
(Hint: Show that there is a compact neighbourhood U of e such that $U \supset K$. Put $X =$
the subgroup generated algebraically by U.)

7. Let G and H be topological groups, and let η be
a homomorphism of H onto the group of automorphisms of G.
We define a group structure on the set $G \times H$ by putting

$$(g_1,h_1) \cdot (g_2,h_2) = (g_1 \eta(h_1)(g_2), h_1 h_2) .$$

Further, let $(g,h) \to \eta(h)(g)$ be a continuous map of $G \times H$
onto G. Show that

      (i)   Each $\eta(h)$ is a homeomorphism of G onto itself

and   (ii)   With the product topology and this group
           structure $G \times H$ is a topological group. (It
           is called the *semidirect product of* G *by* H
           that is determined by η and is denoted by
           $G \times_\eta H$.)

8. (i)   Let G be a σ-compact locally compact Hausdorff
           topological group with N a closed normal sub-
           group of G and H a closed subgroup of G
           such that $G = NH$ and $N \cap H = \{e\}$. Show
           that G is topologically isomorphic to an
           appropriately defined semidirect product
           $N \times_\eta H$.
           (Hint: Let $\eta(h)(n) = h^{-1}nh$, $h \in H$ and $n \in N$.)
    (ii) If H is also normal, show that G is topo-
           logically isomorphic to $N \times H$.
  (iii) If A and B are closed compactly generated
           subgroups of a locally compact Hausdorff abelian

topological group G such that $A \cap B = \{e\}$
and $G = AB$, show that G is topologically
isomorphic to $A \times B$.

9. Let G and H be Hausdorff topological groups and
f a continuous homomorphism of G into H . If G has a
neighbourhood U of e such that $\bar{U}$ is compact and $f(U)$
is a neighbourhood of e in H , show that f is an open
map.

* * * * * * * * *

2 · Subgroups and quotient groups of R^n

In this chapter we expose the structure of the closed subgroups and Hausdorff quotient groups of R^n , $n \geqslant 1$.

Notation. Henceforth we shall focus our attention on abelian groups which will in future be written additively. However, we shall still refer to the product of two groups A and B (and denote it by A × B) rather than the sum of the two groups. We shall also use A^n to denote the product of n copies of A and $\underset{i \in I}{\Pi} A_i$ for the product of the groups A_i , i ∈ I . The identity of an abelian group will be denoted by 0 .

Proposition 19. *Every non-discrete subgroup G of R is dense.*

Proof. We have to show that for each x ∈ R and each ε > 0 , there exists an element g ∈ G ∩ $[x-\varepsilon, x+\varepsilon]$.

As G is not discrete, 0 is not an isolated point. So there exists an element x_ε ∈ (G \ {0}) ∩ $[0, \varepsilon]$. Then the intervals $[nx_\varepsilon, (n+1)x_\varepsilon]$, n = 0,±1,±2,... cover R and are of length $\leqslant \varepsilon$. So for some n , nx_ε ∈ $[x-\varepsilon, x+\varepsilon]$ and of course nx_ε ∈ G . //

Proposition 20. *Let G be a closed subgroup of R . Then G = {0} , G = R , or G is a discrete group of the form aZ = {0,a,-a,2a,-2a,...} , for some a > 0 .*

Proof. Assume G ≠ R . As G is closed, and hence not

27

dense in R , G must be discrete. If G ≠ {0} , then G
contains some positive real number b . So $[0,b] \cap G$ is
a closed non-empty subset of the compact set $[0,b]$. Thus
$[0,b] \cap G$ is compact and discrete. Hence $[0,b] \cap G$ is
finite, and so there exists a least element a > 0 in G .
    For each x ∈ G , let $\left[\frac{x}{a}\right]$ denote the integer part of $\frac{x}{a}$.
Then $x - \left[\frac{x}{a}\right]a \in G$ and $0 \leqslant x - \left[\frac{x}{a}\right]a < a$. So $x - \left[\frac{x}{a}\right]a = 0$;
that is, x = na , for some n ∈ Z , as required. //

Corollary 1. *If a,b ∈ R then gp{a,b} , the subgroup of
R generated by {a,b} , is closed if and only if a and b
are rationally dependent.*

Proof. Exercise.

Examples. gp{1,√2} and gp{√2,√3} are dense in R .

Corollary 2. *Every proper Hausdorff quotient group of R
is topologically isomorphic to T .*

Proof. If R/G is a proper Hausdorff quotient group of
R , then, by Proposition 14, G is a closed subgroup of R .
By Proposition 20, G is of the form aZ , a > 0 . Noting
that the map $x \to \frac{1}{a}x$ is a topological group isomorphism of
R onto itself such that aZ maps to Z , we see that R/aZ
is topologically isomorphic to R/Z which, we know, is topologically isomorphic to T . //

Corollary 3. *Every proper closed subgroup of T is finite.*

Proof. Identify T with the quotient group R/Z and let
p: R → R/Z be the canonical quotient homomorphism. If G
is any proper closed subgroup of R/Z then $p^{-1}(G)$ is a
proper closed subgroup of R . So $p^{-1}(G)$ is discrete. By

Proposition 10, the restriction $p: p^{-1}(G) \to G$ is an open map, so we see that G is discrete. As G is also compact, it is finite. //

We now proceed to the investigation of closed subgroups of R^n, for $n \geq 1$. Here we use the fact that R^n is a vector space over the real field.

Notation: If A is a subset of R^n we denote by $sp_R(A)$ the subgroup $\{\alpha_1 a_1 + \ldots + \alpha_m a_m : \alpha_i \in R, a_i \in A, i = 1, \ldots, m,$ m a positive integer$\}$; and by $sp_Q(A)$ the subgroup $\{\alpha_1 a_1 + \ldots + \alpha_m a_m : \alpha_i \in Q, a_i \in A, i = 1, \ldots, m, m$ a positive integer$\}$; and by $gp(A)$ the subgroup of R^n generated by A.

Clearly $gp(A) \subseteq sp_Q(A) \subseteq sp_R(A)$. We define rank (A) to be the dimension of the vector space $sp_R(A)$.

Proposition 21. *If* $\{a_1, \ldots, a_m\}$ *is a linearly independent subset of* R^n, *then* $gp\{a_1, \ldots, a_m\}$ *is topologically isomorphic to* Z^m.

Proof. Choose elements $a_{m+1}, \ldots, a_n$ so that $\{a_1, \ldots, a_m, a_{m+1}, \ldots, a_n\}$ is a basis for R^n. It is clear that if $\{c_1, \ldots, c_n\}$ is the canonical basis for R^n, then $gp\{c_1, \ldots, c_m\}$ is topologically isomorphic to Z^m. By Problem 2 of Exercise Set Five, every linear transformation of R^n onto itself is a homeomorphism. So the linear map taking a_i to c_i, $i = 1, \ldots, n$, yields a topological group isomorphism of $gp\{a_1, \ldots, a_m\}$ onto $gp\{c_1, \ldots, c_m\}$ $= Z^m$. //

Proposition 22. *Let* G *be a discrete subgroup of* R^n *of rank* p, *and* $a_1, \ldots, a_p \in G$ *a basis for* $sp_R(G)$. *Let* P *be the closed parallelotope with centre* 0 *and basis vectors,* $a_1, \ldots, a_p$; *that is,* $P = \{\sum_{i=1}^{p} r_i a_i : -1 \leq r_i \leq 1,$

$i = 1,\ldots,p$ }. *Then* $G \cap P$ *is finite and* $gp(G \cap P) = G$. *Further, every point in* G *is a linear combination of* $\{a_1,\ldots,a_p\}$ *with rational coefficients; that is* $G \subseteq sp_Q\{a_1,\ldots,a_p\}$.

Proof. As P is compact and G is discrete (and closed in R^n), $G \cap P$ is discrete and compact, and hence finite.

Now $G \subseteq sp_R\{a_1,\ldots,a_p\}$ implies that each $x \in G$ can be written as $x = \sum_{i=1}^{p} t_i a_i$, $t_i \in R$. For each positive integer m , the point

$$z_m = mx - \sum_{i=1}^{p} [mt_i]a_i = \sum_{i=1}^{p} (mt_i - [mt_i])a_i$$

where $[\,]$ denotes "integer part of", belongs to G . As $0 \le mt_i - [mt_i] < 1$, $z_m \in P$. Hence $x = z_1 + \sum_{i=1}^{p} [t_i]a_i$, which says that $gp(G \cap P) = G$.

Further, as $G \cap P$ is finite there exist integers h and k such that $z_h = z_k$. So $(h-k)t_i = [ht_i] - [kt_i]$, $i = 1,\ldots,p$. So $t_i \in Q$, $i = 1,\ldots,p$. Thus $x \in sp_Q\{a_1,\ldots,a_p\}$. //

Corollary. *Let* $\{a_1,\ldots,a_p\}$ *be a linearly independent subset of* R^n , *and* $b = \sum_{i=1}^{p} t_i a_i$, $t_i \in R$. *Then* $gp\{a_1,\ldots,a_p,b\}$ *is discrete if and only if* $t_1,\ldots,t_p$ *are rational numbers.*

Proof. Exercise.

Theorem 4. *Every discrete subgroup* G *of* R^n *of rank* p *is generated by* p *linearly independent vectors, and hence is topologically isomorphic to* Z^p .

Proof. Since G is of rank p , $G \subseteq sp_R\{a_1,\ldots,a_p\}$, where $a_1,\ldots,a_p$ are linearly independent elements of G .

By Proposition 22, $G = gp\{g_1,\ldots,g_r\}$ where each $g_i \in$ $sp_Q\{a_1,\ldots,a_p\}$. So there exists a $d \in Z$ such that $g_i \in gp\{\frac{1}{d} a_1,\ldots, \frac{1}{d} a_p\}$, $i = 1,\ldots,r$.

Now if $\{b_1,\ldots,b_p\}$ is a linearly independent subset of G , then $b_i = \Sigma \beta_{ij} a_j$, where the determinant, $\det(\beta_{ij}) \neq 0$, and $\beta_{ij} \in \frac{1}{d} Z$. So $\det(\beta_{ij}) \in \frac{1}{d^p} Z$. So out of all such $\{b_1,\ldots,b_p\}$ there exists one with $|\det(\beta_{ij})|$ minimal. Let this set be denoted by $\{b_1,\ldots,b_p\}$. We claim that $G = gp\{b_1,\ldots,b_p\}$ and hence is topologically isomorphic to Z^p .

Suppose $G \neq gp\{b_1,\ldots,b_p\}$. Then there exists an element $g \in G$ with $g = \sum_{i=1}^{p} \lambda_i b_i$ and not all $\lambda_i \in Z$. Suppose that $\lambda_1 = \frac{r}{s}$, $r \neq 0$ and $s > 1$. Since $b_1 \in G$ we can assume that $|\lambda_1| < 1$ (by subtracting multiples of b_1 , if necessary). Then putting $b_1' = g$, $b_i' = b_i$, $i = 2,\ldots,p$ and $b_i' = \Sigma \beta_{ij}' a_j$ we see that

$$\det(\beta_{ij}') = \det \begin{bmatrix} \lambda_1 & & & & \\ \lambda_2 & 1 & & O & \\ \vdots & & 1 & & \\ & & & \ddots & \\ \lambda_p & & O & & 1 \end{bmatrix} \det(\beta_{ij}) = \lambda_1 \det(\beta_{ij}) .$$

As $|\lambda_1| < 1$ this means that $|\det(\beta_{ij}')| < |\det(\beta_{ij})|$, which is a contradiction. //

Proposition 23. *Every non-discrete closed subgroup* H *of* R^n , $n \geqslant 1$, *contains a line through zero.*

Proof. As H is non-discrete there exists a sequence $h_1, h_2, \ldots$ of points in H converging to 0 , with each $h_n \neq 0$. Let C be an open cube with centre 0 containing all the h_n . Let m_n denote the largest integer $m > 0$ such that $mh_n \in C$. The points $m_n h_n$, $n = 1,2,\ldots$ lie in a compact set $\bar{C}$ and therefore have a cluster point

$a \in \bar{C} \cap H$.

If $\|m_n h_n - a\| \leq \varepsilon$ we have $\|(m_n + 1)h_n - a\| \leq \varepsilon + \|h_n\|$, where $\|\ \|$ denotes the usual norm in R^n . Since $h_n \to 0$ as $n \to \infty$ it follows that a is also a cluster point of the sequence $(m_n + 1)h_n$, $n = 1,2,\ldots$, whose points belong to the closed set $R^n \backslash C$. Hence $a \in \bar{C} \cap (R^n \backslash C)$ – the boundary of C , which implies $a \neq 0$.

Let t be any real number. Since $\left| tm_n - \left[tm_n \right] \right| < 1$, the relation $\|m_n h_n - a\| \leq \varepsilon$ implies that $\| \left[tm_n \right] h_n - ta \|$ $\leq |t|\varepsilon + \|h_n\|$; since $h_n \to 0$ as $n \to \infty$, ta is a limit point of the sequence $\left[tm_n \right] h_n$, $n = 1,2,\ldots$ But the points of this sequence belong to H and so $ta \in H$, since H is closed. So H contains the line through $a \neq 0$ and 0 . //

Theorem 5. *Let* G *be a closed subgroup of* R^n , $n \geqslant 1$. *Then there are (closed) vector subspaces* U , V *and* W *of* R^n *such that*

(i) $R^n = U \times V \times W$

(ii) $G \cap U = U$

(iii) $G \cap V$ *is discrete*

(iv) $G \cap W = \{0\}$

(v) $G = (G \cap U) \times (G \cap V)$.

Proof. Let U be the union of all lines through 0 lying entirely in G . We claim that U is a vector subspace of R^n . To see this let x and y be in U and λ, μ and $\delta \in R$. Then $\delta\lambda x$ is in U and hence also in G . Similarly $\delta\mu y \in G$. So $\delta(\lambda x + \mu y) = \delta\lambda x + \delta\mu y \in G$. As this is true for all $\delta \in R$, we have that $\lambda x + \mu y \in U$. So U is a vector subspace of R^n , and $G \cap U = U$.

Let U' be any complementary subspace of U ; that is, $R^n = U \times U'$. So if $g \in G$, then $g = h + k$, $h \in U$, $k \in U'$. As $G \supseteq U$, $h \in G$ so $k = g - h \in G$. Hence

$$G = U \times (G \cap U') \ .$$

Put $V = \mathrm{sp}_R(G \cap U')$ and $W =$ a complementary subspace in U' of V . So $G \cap W = \{0\}$. Clearly $G \cap V$ contains no lines through 0 , which by Proposition 23, implies that $G \cap V$ is discrete. //

Theorem 6. *Let G be a closed subgroup of R^n , $n \geqslant 1$.*
If r equals the rank of G (that is, $\mathrm{sp}_R(G)$ has dimension r) then there exists a basis $a_1,\ldots,a_n$ of R^n such that

$$G = \mathrm{sp}_R\{a_1,\ldots,a_p\} \times \mathrm{gp}\{a_{p+1},\ldots,a_r\} \ .$$

So G is topologically isomorphic to $R^p \times Z^{r-p}$ and the quotient group R^n/G is topologically isomorphic to $T^{r-p} \times R^{n-r}$.

Before stating the next theorem let us record some facts about free abelian groups.

Definition. A group F is said to be a *free abelian group* if it is the restricted direct product of a finite or infinite number of infinite cyclic groups. Each of these infinite cyclic groups has a single generator and the set S of these generators is said to be a *basis* of F .

Remarks.
(i)   It can be shown that an abelian group F is a free abelian group with basis S if and only if S is a subset of F with the property that every map f of S into any abelian group G can be extended uniquely to a homomorphism of F into G .
(ii)   One consequence of (i) is that any abelian group G is a quotient group of some free abelian group. (Let F be the free abelian group with basis S of the same cardinality as G . Then there is a bijection

ϕ of S onto G . Extend this map to a homomorphism Φ of F onto G .)

(iii) Proposition 21 together with Theorem 4 show that any subgroup of Z^n is isomorphic to Z^m , for some m . In other words, any subgroup of a free abelian group with finite basis is a free abelian group with finite basis. It can be shown that any subgroup of a free abelian group is a free abelian group. For details see A.G. Kurosh, "The theory of groups I", pp.142-144.

(iv) Finally, we record that if the abelian group G admits a homomorphism ϕ onto a free abelian group F then G is isomorphic to F × A , where A is the kernel of ϕ . (Note that it suffices to produce a homomorphism θ of F into G such that $\phi\theta$ is the identity map of F . To produce θ , let S be a basis of F and for each $s \in S$ choose a $g_s \in G$ such that $\phi(g_s) = s$. As F is a free abelian group the map $s \to g_s$ of S into G can be extended to a homomorphism θ of F into G . Clearly $\phi\theta$ acts identically on F .)

Theorem 7. *Let* H = V × F , *where* V *is a divisible abelian Hausdorff group and* F *is a discrete free abelian group. If* G *is a closed subgroup of* H , *then there exists a discrete free abelian subgroup* F' *of* H *isomorphic to* F *such that*

(i) H = V × F'

and (ii) G = (G ∩ V) × (G ∩ F') .

Proof. Let π_1: H → V and π_2: H → F be the projections. The restriction of π_2 to G is a homomorphism from G to F with kernel G ∩ V . Since F is a free abelian group, and every subgroup of a free abelian group is a free abelian group, G/(G ∩ V) is free abelian, and therefore, by the

34

above Remark (iv), G is algebraically isomorphic to $(G \cap V) \times C$, where C is a free abelian subgroup of G.

Let p_1 and p_2 be the restrictions of π_1 and π_2 to C, respectively. Then p_2 is one-one as $C \cap V = C \cap G \cap V = \{0\}$.

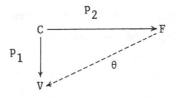

We can define a homomorphism $\theta: p_2(C) \to V$ by putting $\theta(p_2(c)) = p_1(c)$ and then use Proposition 17 to extend θ to a homomorphism of F into the divisible group V. So $\theta p_2 = p_1$. If we now define a homomorphism $\phi: F \to H$ by $\phi(x) = \theta(x) + x$ and put $F' = \phi(F)$ we have that $H = V \times F'$, algebraically; the decomposition being given by

$$v + f = \left[v - \theta(f) \right] + \left[\theta(f) + f \right], \quad v \in V \text{ and } f \in F.$$

Also $C \subseteq F'$, since for each c in C we have

$$c = p_1(c) + p_2(c) = \theta(p_2(c)) + p_2(c) = \phi(p_2(c)) \in \phi(F) = F'.$$

So (i) and (ii) are satisfied algebraically.

Now $\phi: F \to F'$ is an algebraic isomorphism and since ϕ^{-1} is induced by π_2, ϕ^{-1} is continuous. But F is discrete, so ϕ is a homeomorphism and F' is a discrete free abelian group.

To show that H has the product topology with respect to the decomposition $H = V \times F'$, it suffices to show that the corresponding projections $\pi_1': H \to V$ and $\pi_2': H \to F'$ are continuous. But this is clearly the case since $\pi_1'(h) = \pi_1(h) - \theta(\pi_2(h))$ and $\pi_2'(h) = \pi_2(h) + \theta(\pi_2(h))$, for each

$h \in H$. Hence the decomposition $G = (G \cap V) \times (G \cap F')$ also has the product topology. //

Corollary 1. *Let* G *be a closed subgroup of* $R^n \times Z^m$. *Then* G *is topologically isomorphic to* $R^a \times Z^b$, *where* $a \leqslant n$ *and* $a + b \leqslant n + m$. *Further* $(R^n \times Z^m)/G$ *is topologically isomorphic to* $R^c \times T^d \times D$, *where* D *is a discrete finitely generated abelian group (with* $f \leqslant m$ *generators) and* $c + d \leqslant n$.

Corollary 2. *Let* G *be a closed subgroup of* $R^n \times T^m \times D$, *where* D *is a discrete abelian group. Then* G *is topologically isomorphic to* $R^a \times T^b \times D'$, *where* D' *is a discrete group* $a + b \leqslant n + m$. *Further* $(R^n \times T^m \times D)/G$ *is topologically isomorphic to* $R^c \times T^d \times D''$, *where* D'' *is a discrete group and* $c + d \leqslant n + m$.

Proof. Let F be a discrete free abelian group with D as a quotient group. (See the Remarks preceding Theorem 7.) Then there is a natural quotient homomorphism p of $R^{n+m} \times F$ onto $R^n \times T^m \times D$. Then G is a quotient group of $p^{-1}(G) \leqslant R^{n+m} \times F$. Now Theorem 7 together with Theorem 6 describe both $p^{-1}(G)$ and the kernel of the map of $p^{-1}(G)$ onto G , and yield the result. //

Remark. In Corollary 2 we have not said that $a \leqslant n$, $b \leqslant m$ and $c \leqslant n$. These inequalities are indeed true. They will follow from the above once we have the Pontryagin-van Kampen duality theorem.

Corollary 3. *Let* G *be a closed subgroup of* T^n . *Then* G *is topologically isomorphic to* $T^a \times D$ *where* D *is a finite discrete group and* $a \leqslant n$.

Proof. Exercise.

Definition. The topological groups G and H are said to
be *locally isomorphic* if there are neighbourhoods V of e
in G and U of e in H and a homeomorphism f of V
onto U such that if x , y and xy all belong to V
then f(xy) = f(x)f(y) .

Example. R and T are locally isomorphic.

Proposition 24. *If D is a discrete normal subgroup of a*
topological group G , then G and G/D are locally iso-
morphic.

Proof. Exercise.

Lemma. Let U be a neighbourhood of 0 in an abelian
topological group G and V be a neighbourhood of 0 in
R^n , n ⩾ 1 . If there is a map f of V onto U such
that x ∈ V , y ∈ V and x + y ∈ V implies f(x+y) = f(x)
+ f(y) , then f can be extended to a continuous homomorphism
of R^n onto the open subgroup of G generated by U .

Proof. Exercise.

Theorem 8. *Let G be a Hausdorff abelian topological group*
locally isomorphic to R^n , n ⩾ 1 . Then G is topologically
isomorphic to $R^a \times T^b \times D$, where D is a discrete group
and a + b = n .

Proof. By the above Lemma there is a continuous homomorphism
f of R^n onto an open subgroup H of G . As G is locally
isomorphic to R^n , it has a compact neighbourhood of 0 and
so is locally compact. Hence H is locally compact and the

37

Open Mapping Theorem (Theorem 3) says that f is an open map; that is, H is a quotient group of R^n . Further the kernel K of f is discrete since otherwise there would be elements $x \neq 0$ of K arbitrarily close to 0 such that $f(x) = 0$, which is false as f maps a neighbourhood of 0 homeomorphically into G . So Theorem 6 tells us that H is topologically isomorphic to $R^a \times T^b$, with $a + b = n$.

Now H is an open divisible subgroup of G which, by Proposition 18, implies that G is topologically isomorphic to $H \times D$, where $D = G/H$ is discrete. Thus G is topologically isomorphic to $R^a \times T^b \times D$, as required. //

Corollary. *Any connected topological group locally isomorphic to* R^n *,* $n \geqslant 1$ *, is topologically isomorphic to* $R^a \times T^b$ *, where* $a + b = n$ *.*

Remark. We conclude this chapter by noting that some of the results presented here can be extended from finite to infinite products of copies of R . For example, it is known that any closed subgroup of a countable product $\prod_{i=1}^{\infty} R_i$ of isomorphic copies R_i of R is topologically isomorphic to a countable product of isomorphic copies of R and Z . However, this result does not extend to uncountable products. For details, see R. Brown, P.J. Higgins and S.A. Morris, "Countable products and sums of lines and circles: their closed subgroups, quotients and duality properties", *Math. Proc. Camb. Philos. Soc.* 78 (1975), 19-32; and H. Leptin, "Zur Dualitätstheorie projectiver Limites abelscher Gruppen", *Abh. Math. Sem. Univ. Hamburg* 19 (1955), 264-268.

EXERCISE SET FIVE
    1. If $a,b \in R$ show that the subgroup of R generated

by {a,b} is closed if and only if a and b are rationally dependent.

2. Prove that any linear transformation of the vector space R^n , $n \geqslant 1$, onto itself is a homeomorphism .

3. (i) Let $\{a_1, \ldots, a_p\}$ be a linearly independent subset of R^n , $n \geqslant 1$, and $b = \sum_{i=1}^{p} t_i a_i$, $t_i \in R$. Show that $gp\{a_1, \ldots, a_p, b\}$ is discrete if and only if $t_1, \ldots, t_p$ are rational numbers.

(ii) Hence prove the following (diophantine approximation) result: Let $\theta_1, \ldots, \theta_n$ be n real numbers. In order that for each $\varepsilon > 0$ there exist an integer q and n integers p_i , $i = 1, \ldots, n$ such that

$$|q\theta_i - p_i| \leqslant \varepsilon \ , \quad i = 1, \ldots, n$$

where the left hand side of at least one of these inequalities does not vanish, it is necessary and sufficient that at least one of the θ_i be irrational.

4. Show that if G and H are locally isomorphic topological groups then there exists a neighbourhood V' of e in G and U' of e in H and a homeomorphism f of V' onto U' such that if x , y and xy all belong to V' then $f(xy) = f(x)f(y)$ and if x' , y' and x'y' all belong to U' then $f^{-1}(x'y') = f^{-1}(x')f^{-1}(y')$.

5. (i) Verify that any topological group locally isomorphic to a Hausdorff topological group is Hausdorff.

(ii)   Verify that any connected topological group
       locally isomorphic to an abelian topological
       group is abelian.

(iii) Deduce that any connected topological group
       locally isomorphic to R^n , $n \geqslant 1$, is topo-
       logically isomorphic to $R^a \times T^b$, where
       $a + b = n$.

6. Show that if D is a discrete normal subgroup of a
topological group G , then G and G/D are locally isomorphic.

7. Let U be a neighbourhood of 0 in an abelian topological group and V a neighbourhood of 0 in R^n , $n \geqslant 1$.
If there is a map f of V onto U such that $x \in V$,
$y \in V$ and $x + y \in V$ implies $f(x + y) = f(x) + f(y)$, show
that f can be extended to a continuous homomorphism of R^n
onto the open subgroup of G generated by U .

* * * * * * * * *

3 · Uniform spaces and dual groups

Uniform Spaces. We now say a very few words about uniform
spaces. For further discussion, see Kelley [General Topology]
and Bourbaki.

We introduce some notation convenient for this discussion.
Let X be a set and $X \times X = X^2$ the product of X with
itself. If V is a subset of X^2 then V^{-1} denotes the
set $\{(y,x): (x,y) \in V\} \subseteq X^2$. If U and V are subsets
of X^2 then UV denotes the set of all pairs (x,z) , such
that for some $y \in X$, $(x,y) \in U$ and $(y,z) \in V$. Putting
$V = U$ defines U^2 . The set $\{(x,x): x \in X\}$ is called the
diagonal.

Definition. A *uniformity* on a set X is a non-void family
U of subsets of $X \times X$ such that
  (a) Each member of U contains the diagonal
  (b) $U \in U \Rightarrow U^{-1} \in U$
  (c) if $U \in U$ then there is a $V \in U$ such that $V^2 \subseteq U$
  (d) if $U \in U$ and $V \in U$, then $U \cap V \in U$
  (e) if $U \in U$ and $U \subseteq V \subseteq X^2$, then $V \in U$.
The pair (X,U) is called a *uniform space*.

Examples. If R is the set of real numbers then the "usual
uniformity" for R is the family U of all subsets U of
$R \times R$ such that $\{(x,y): |x-y| < r\} \subseteq U$, for some positive
real number r .

Indeed if (X,d) is any metric space then we can define
a uniformity U on X by putting U equal to the collection
of all subsets U of $X \times X$ such that $\{(x,y): d(x,y) < r\} \subseteq U$,

for some positive real number r .

Let (G,τ) be a topological group and for each neigh-bourhood U of e let $U_L = \{(x,y): x^{-1}y \in U\}$ and
$U_R = \{(x,y): xy^{-1} \in U\}$. Then the *left uniformity* L on
G consists of all sets $V \subseteq G \times G$ such that $V \supseteq U_L$,
for some U . Similarly we define the *right uniformity*.
The two-sided uniformity consists of all sets W such that
$W \supseteq U_L$ or $W \supseteq U_R$, for some U .

Given any uniformity U on a set X we can define a
corresponding topology on X . For each $x \in X$, let
$U_x = \{y \in X: (x,y) \in U\}$. Then as U runs over U , the
system U_x defines a base of neighbourhoods at x for a
topology; that is, a subset T of X is open in the topo-logy if and only if for each $x \in T$ there is a $U \in U$ such
that $U_x \subseteq T$.

It is easily verified that if (G,τ) is a topological
group then the topologies arising from the left uniformity,
the right uniformity and the two-sided uniformity all
agree with the given topology.

Definition. Let E and F be topological spaces and M
any collection of subsets M of E and {V} a base of
open sets in F . Let $P(M,V) = \{f: f \in F^E$ and $f(M) \subseteq V\}$.
(F^E denotes the collection of functions $f: E \to F$.) The
family $\{P(M,V)\}$, where M runs over M and V runs over
{V} , is a subbase for a topology on F^E .

If F is a Hausdorff space and M is a covering of E
then it is easily verified that this topology is Hausdorff.
Two important special cases of this topology are when

    (a) M is the collection of all finite subsets of E
          - the topology is then the *p-topology* or *the topology*
          of pointwise convergence, and

    (b) when M is the collection of all compact subsets
          of E - the *k-topology* or the *compact open topology*.

Since every finite set is compact, $k \supseteq p$. Therefore a subset of F^E which is k-compact is also p-compact, but the converse is false.

Observe that F^E with the p-topology is simply $\prod_{x \in E} F_x$, with the product topology, where each F_x is a homeomorphic copy of F .

We are interested in $C(E,F)$ the subset of F^E consisting of all *continuous* functions from E to F , and we shall want to find conditions which guarantee that a subset of $C(E,F)$ is k-compact.

Definition. Let E and F be topological spaces and G a subset of F^E . A topology on G is said to be *jointly continuous* if the map θ from the product space $G \times E$ to F , given by $\theta(g,x) = g(x)$, is continuous.

Proposition 25. *Each topology on* $G \subseteq F^E$ *which is jointly continuous is finer than the* k-*topology.*

Proof. Let τ be a jointly continuous topology on G , U an open subset of F , K a compact subset of E and θ the map taking (g,x) to $g(x)$, $g \in G$ and $x \in E$. We want to show that for each $f \in P(K,U) = \{g: g \in G$ and $g(K) \subseteq U\}$ there is a set $W \in \tau$ such that $f \in W \subseteq P(K,U)$.

As θ is jointly continuous the set $V = (G \times K) \cap \theta^{-1}(U)$ is open in $G \times K$. If $f \in P(K,U)$ then $\{f\} \times K \subseteq V$ and since $\{f\} \times K$ is compact, there is a $W \in \tau$ such that $f \in W$ and $W \times K \subseteq \theta^{-1}(U)$. Hence $W \subseteq P(K,U)$ as required. //

Proposition 26. *Let* E *and* F *be topological spaces and* $G \subseteq C(E,F)$. *Then* G *is* k-*compact if*

    (a) G *is* k-*closed in* $C(E,F)$

    (b) *the closure of the set* $\{g(x): g \in G\}$ *is compact, for each* $x \in E$

and (c) *the* p-*topology for the* p-*closure of* G *in* F^E *is jointly continuous.*

Proof. Let $\bar{G}$ be the p-closure in F^E of G. By condition (b), $\underset{x \in E}{\Pi} \overline{(\{g(x): g \in G\})}$ is a p-compact set, and since $\bar{G}$ is a p-closed subset of this set, $\bar{G}$ is p-compact. By condition (c), the p-topology on $\bar{G}$ is jointly continuous - so $\bar{G} \subseteq C(E,F)$. Also by Proposition 25, the p-topology on $\bar{G}$ is finer than the k-topology and hence they coincide. Thus $\bar{G}$ is k-compact. As G is k-closed in $C(E,F)$ and $\bar{G}$ is k-compact and a subset of $C(E,F)$, we have that G is k-compact. //

Definition. Let E be a topological space and F a uniform space. A subset G of $C(E,F)$ is said to be *equicontinuous at the point* $x \in E$ if for each U in the uniformity U of F, there exists a neighbourhood V of x such that $(g(y),g(x)) \in U$, for all $y \in V$ and $g \in G$. The family G is said to be *equicontinuous* if it is equicontinuous at every $x \in E$.

Proposition 27. *Let* G *be a subset of* $C(E,F)$ *which is equicontinuous at* $x \in E$. *Then the* p-*closure* $\bar{G}$ *in* F^E *of* G *is also equicontinuous at* x.

Proof. Exercise.

Proposition 28. *Let* G *be an equicontinuous subset of* $C(E,F)$. *Then the* p-*topology on* G *is jointly continuous.*

Proof. Exercise.

By combining Propositions 26, 27 and 28 we obtain the

following:

Theorem 9 (Ascoli's Theorem). *Let* E *be a topological*
space and F *a uniform space. A subset* G *of* $C(E,F)$ *is*
k-compact if

    (a)   G *is* k-*closed in* $C(E,F)$

    (b)   *the closure of the set* $\{g(x): g \in G\}$ *is compact,*
        for each $x \in E$

and (c)   G *is equicontinuous.*

Remark.   If E is a locally compact Hausdorff space and
F is a Hausdorff uniform space then the converse of Theorem
9 is valid; that is, any k-compact subset G of $C(E,F)$
satisfies conditions (a), (b) and (c). (See Kelley, General
Topology.)

EXERCISE SET SIX

   1. Let (X,U) be any uniform space and (X,τ) the
associated topological space. Show that (X,τ) is a regular
space.

   2. If (G,τ) is a topological group show that the topologies associated with the left uniformity on G , the right
uniformity on G , and the two-sided uniformity on G coincide with τ .

   3. (i)   Let G be a topological group and $\{U_n: n =$
            $1,2,\ldots\}$ a base for the left uniformity on G
            such that

          (a) $\bigcap\limits_{n=1}^{\infty} U_n$ = diagonal of $G \times G$

          (b) $U_{n+1}U_{n+1}U_{n+1} \subseteq U_n$

(c) $U_n = U_n^{-1}$, for each n .

Show that there exists a metric d on G such
that $U_n \subseteq \{(x,y): d(x,y) < 2^{-n}\} \subseteq U_{n-1}$, for
each $n > 1$.
(Hint: Define a real-valued function f on
$G \times G$ by letting $f(x,y) = 2^{-n}$ if $(x,y) \in$
$U_{n-1} \backslash U_n$ and $f(x,y) = 0$ if (x,y) belongs to
each U_n . The desired metric d is constructed
from its "first approximation", f , by a chaining
argument. For each x and y in G let
$d(x,y)$ be the infimum of $\{\sum_{i=0}^{n} f(x_i,x_{i+1})\}$
over all finite sequences $x_0, x_1, \ldots, x_{n+1}$ such
that $x = x_0$ and $y = x_{n+1}$.)

(ii)   Prove that a topological group is metrizable if
and only if it satisfies the first axiom of
countability at the identity; that is, there is
a countable base of neighbourhoods at the
identity.

4. Let E be a topological space, F a uniform space
and G a subset of $C(E,F)$. Show that
(i)   if G is equicontinuous at $x \in E$, then the
p-closure in F^E of G is also equicontinuous
at x
and (ii)   if G is an equicontinuous subset of $C(E,F)$,
then the p-topology on G is jointly continuous.

* * * * * * * * *

We are now ready to begin our study of duality.

Definitions. If G is an abelian topological group then a continuous homomorphism $\gamma: G \to T$ is said to be a *character*. The collection of all characters is called the *character group* or *dual group* of G, and is denoted by G^* or Γ.

Observe that G^* is an abelian group if for each γ_1 and γ_2 in G^* we define

$$(\gamma_1 + \gamma_2)(g) = \gamma_1(g) + \gamma_2(g) , \quad \text{for all } g \in G .$$

Instead of writing $\gamma(g)$, $\gamma \in \Gamma$ and $g \in G$ we shall generally write (g, γ).

Example 1. Consider the group Z. Each character γ of Z is determined by $\gamma(1)$, as $\gamma(n) = n\gamma(1)$, for each $n \in Z$. Of course $\gamma(1)$ can be any element of T. For each $a \in T$, let γ_a denote the character γ of Z with $\gamma(1) = a$. Then the mapping $a \to \gamma_a$ is clearly an algebraic isomorphism of T onto the character group of Z. So the dual group Z^* of Z is algebraically isomorphic to T.

Example 2. Consider the group T. We claim that every character γ of T can be expressed in the form $\gamma(x) = mx$, where m is an integer characterizing the homomorphism γ.

To see this let K denote the kernel of γ. Then by Corollary 3 of Proposition 20, $K = T$ or K is a finite cyclic group. If $K = T$, then γ is the trivial character and $\gamma(x) = 0.x$, $x \in T$. If K is a finite cyclic group of order r then, by Corollary 2 of Proposition 20, T/K is topologically isomorphic to T. Indeed, if p is the canonical map of T onto T/K then the topological isomorphism $\theta: T/K \to T$ is such that $\theta p(x) = rx$. Let α the continuous one-one homomorphism of T into T induced by γ.

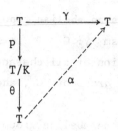

Problem 1 of Exercise Set Seven implies that $\alpha(x) = x$, for all $x \in T$, or $\alpha(x) = -x$, for all $x \in T$. So $\gamma(x) = rx$ or $-rx$, for each $x \in T$.

Hence each character γ of T is of the form $\gamma = \gamma_m$ for some $m \in Z$, where $\gamma_m(x) = mx$ for all $x \in T$. Of course $\gamma_m + \gamma_n = \gamma_{m+n}$. Thus the dual group T^* of T is algebraically isomorphic to Z , with the isomorphism being $m \to \gamma_m$.

Example 3.   Consider the group R . We claim that every character γ of R can be expressed in the form $\gamma(x) =$ $\exp(2\pi i dx)$, $x \in R$, where d is a fixed real number defining γ : $\gamma = \gamma_d$. Further, we have $\gamma_a + \gamma_b = \gamma_{a+b}$. Thus the dual group of R is algebraically isomorphic to R itself, under the isomorphism $d \to \gamma_d$.

To prove the claim, let K denote the kernel of γ . If $K = R$ then $\gamma = \gamma_0$. If $K \neq R$ then Proposition 20 says that K is isomorphic to Z . Further by Corollary 2 of Proposition 20 the quotient group R/K is topologically isomorphic to T . As in Example 2 there are only two possibilities for the induced algebraic isomorphism $R/K \to T$;   these give rise to the cases $\gamma = \gamma_{1/a}$ and $\gamma = \gamma_{(-1/a)}$, where a is the least element of K . So every γ is of the form γ_d for some $d \in R$, and hence R is algebraically isomorphic to its dual group.

We now topologize G^* .

Remark. Note that G^* is a p-closed subset of $C(G,T)$.

Proposition 29. *Let* G *be any abelian topological group.*
Then G^* *endowed with the p-topology or the* k-*topology is*
a Hausdorff abelian topological group.

Proof. Exercise.

Theorem 10. *If* G *is any LCA-group then* G^* , *endowed*
with the k-*topology, is an LCA-group.*

Proof. To show that G^* , with the k-topology, is locally
compact, let U be any compact neighbourhood of 0 in G
and $V_a = \{t: t = \exp(2\pi ix) \in T$ and $1 > x > 1-a$ or
$a > x \geqslant 0\}$, where a is a positive real number $< \frac{1}{4}$.
Then V_a is an open neighbourhood of 0 in T .
   Let $N_a = P(U,V_a) = \{\gamma \in G^*: (g,\gamma) \in V_a$, for each
$g \in U\}$. By the definition of the k-topology, N_a is a
neighbourhood of 0 in G^* . We shall show that the k-
closure of N_a , $cl_k(N_a)$, is k-compact. To do this we use
Theorem 9.
   Firstly we show that N_a is equicontinuous. Let $\varepsilon > 0$
be given. We wish to show that there exists a neighbourhood
U_1 of 0 in G such that for all $\gamma \in N_a$ and g , h and
$g - h$ in U_1 , $(g-h,\gamma) = (g,\gamma) - (h,\gamma) \in V_\varepsilon$, where
$V_\varepsilon = \{t: t = \exp(2\pi ix) \in T$ and $1 > x > 1-\varepsilon$ or $\varepsilon > x \geqslant 0\}$.
   Suppose that there is no such U_1 . Without loss of
generality assume $\varepsilon < \frac{1}{4}$ and let n be a positive integer
such that $\frac{1}{2} > n\varepsilon > a$. Further, let W be a neighbourhood
of 0 in G such that

(1) $\sum_{i=1}^{n} W_i \subseteq U$, where each $W_i = W$.

By assumption, then, for some g and h in W with

$g - h \in W$ and some $\gamma \in N_a$, $(g - h, \gamma) \notin V_\varepsilon$. So without
loss of generality $(g - h, \gamma) = \exp(2\pi ix)$ with $a > x \geqslant \varepsilon$.
Let j be a positive integer $\leqslant n$ such that $\frac{1}{2} > jk > a$.
So $(j(g - h), \gamma) \notin V_a$. But as jg , jh and $j(g - h)$ all
belong to U , by (1), $(j(g - h), \gamma) \in V_a$ - which is a
contradiction. Hence N_a is equicontinuous.

By Proposition 27, the p-closure of N_a is equicontinuous.
As any subset of an equicontinuous set is equicontinuous, and
$cl_k(N_a)$ is a subset of the p-closure of N_a , we have that
$cl_k(N_a)$ is equicontinuous.

As T is compact, condition (b) of Theorem 9 is also
satisfied and hence $cl_k(N_a)$ is a compact neighbourhood of
0 . So G^* with the k-topology is locally compact. //

As a corollary to the proof of Theorem 10 we have

Theorem 11. *Let* G *be any LCA-group,* Γ *its dual group
endowed with the* k*-topology,* K *a compact neighbourhood of*
0 *in* G *and* $V_a = \{t: t = \exp(2\pi ix) \in T \text{ with } 1 > x > 1-a$
or $a > x \geqslant 0\}$, *for some positive real number* $a < \frac{1}{4}$.
Then $\overline{P(K, V_a)}$ *is a compact neighbourhood of* 0 *in* Γ .

Notation. From now on G^* and Γ will denote the dual
group of G *with* the k-topology.

Theorem 12. *Let* G *be an LCA-group and* Γ *its dual group.
If* G *is compact then* Γ *is discrete. If* G *is discrete
then* Γ *is compact.*

Proof. Let G be compact and V_a be as in Theorem 11.
Then $P(G, V_a)$ is a neighbourhood of 0 in Γ . As V_a
contains no subgroup other than $\{0\}$, we must have
$P(G, V_a) = \{0\}$. So Γ has the discrete topology.

Let G be discrete. Then by Theorem 11, $\overline{P(\{0\}, V_a)}$ is
a compact subset of Γ . But $\overline{P(\{0\}, V_a)}$ clearly equals Γ ,

and hence Γ is compact. //

Corollary. *The dual group* T^* *of* T *is topologically isomorphic to* Z .

EXERCISE SET SEVEN

1. Show that if γ is a continuous $1 - 1$ homomorphism of T into itself then either $\gamma(x) = x$, for all $x \in T$ or $\gamma(x) = -x$, for all $x \in T$. (Hint: Firstly show that γ must be onto. Next, observe that T has only one element of order 2 .)

2. Show that if G is any abelian topological group, then G^* endowed with the p-topology or the k-topology is a Hausdorff topological group. (Hint: Let $\gamma_1 - \gamma_2 \in P(K,U)$. Let W be an open symmetric neighbourhood of o in T such that $2W + (\gamma_1 - \gamma_2)(K) \subseteq U$. Observe that

$$\left[\gamma_1 + P(K,W)\right] - \left[\gamma_2 + P(K,W)\right] \subseteq P(K,U) \ .)$$

3. Show that the dual group of Z is topologically isomorphic to T .

4. Show that R is topologically isomorphic to its dual group.

5. Find the dual groups of the discrete finite cyclic groups.

6. Let G be any abelian topological group and G^* its dual group. Show that the family of all sets $P(K,V_\varepsilon)$, as K ranges over all compact subsets of G containing 0 and ε ranges over all positive numbers less than one, is

a base of open neighbourhoods of 0 for the k-topology
on G* .

* * * * * * * * *

4 · Introduction to the Pontryagin – van Kampen duality theorem

We begin with a *statement* of the duality theorem.

Theorem. *Let* G *be an LCA-group and* Γ *its dual group.*
For fixed $g \in G$ *, let* g' *be the function* : Γ → T *given*
by $g'(\gamma) = \gamma(g)$ *, for all* $\gamma \in \Gamma$ *. If* α *is the mapping*
given by $\alpha(g) = g'$ *, then* α *is a topological group iso-*
morphism of G *onto* Γ^* *.*

Remarks.

(i)   Roughly speaking this says that every LCA-group is
        the dual group of its dual group.

(ii)   This theorem says that every piece of information
        about an LCA-group is contained in some piece of
        information about its dual group. In particular all
        information about a compact Hausdorff abelian group
        is contained in information about its dual group - a
        discrete abelian group. So any compact Hausdorff
        abelian group can be completely described by the
        purely algebraic properties of its dual group; for
        example, if G is a compact Hausdorff abelian group
        then we shall see that
        (a) G is metrizable if and only if Γ is countable.
        (b) G is connected if and only if Γ is torsion-free.

Lemma.   In the notation of the above Theorem, α is a continuous homomorphism of G into Γ^* .

Proof.   Firstly we have to show that $\alpha(g) \in \Gamma^*$; that is,

$\alpha(g) = g'$ is a continuous homomorphism of Γ into T. As $\alpha(g)(\gamma_1 + \gamma_2) = (\gamma_1 + \gamma_2)(g) = \gamma_1(g) + \gamma_2(g) = \alpha(g)(\gamma_1) + \alpha(g)(\gamma_2)$, for each γ_1 and γ_2 in Γ, $\alpha(g)$ is a homomorphism : $\Gamma \to T$. To see that $\alpha(g)$ is continuous it suffices to note that $\alpha(g)(\gamma) \in V_\varepsilon$ whenever $\gamma \in P(\{g\}, V_\varepsilon)$, where V_ε is an ε-neighbourhood of 0 in T as in Theorem 10. So α is a map of G into Γ^*.

That α is a homomorphism follows by observing $\alpha(g_1 + g_2)(\gamma) = \gamma(g_1 + g_2) = \gamma(g_1) + \gamma(g_2) = \alpha(g_1)(\gamma) + \alpha(g_2)(\gamma)$, for all $\gamma \in \Gamma$, implies that $\alpha(g_1 + g_2) = \alpha(g_1) + \alpha(g_2)$, for all g_1 and g_2 in G.

To show that α is continuous, it suffices to verify continuity at $0 \in G$. Let W be any neighbourhood of 0 in Γ^*. Without loss of generality we can assume $W = P(K, V_\varepsilon)$, for some compact subset K of Γ. We have to find a neighbourhood of 0 in G which maps into W.

Let U be any open neighbourhood of 0 in G such that $\bar{U}$ is compact and consider the neighbourhood $P(\bar{U}, V_{\varepsilon/2})$ of 0 in Γ. The collection $\{\gamma + P(\bar{U}, V_{\varepsilon/2}) : \gamma \in \Gamma\}$ covers the compact set K and so there exist $\gamma_1, \ldots, \gamma_m$ in Γ such that $K \subseteq [\gamma_1 + P(\bar{U}, V_{\varepsilon/2})] \cup \ldots \cup [\gamma_m + P(\bar{U}, V_{\varepsilon/2})]$. Let U_1 be a neighbourhood of 0 in G such that $U_1 \subseteq U$ and $\gamma_i(g) \in V_{\varepsilon/2}$, for all $g \in U_1$ and $i = 1, \ldots, m$. (This is possible since the γ_i are continuous.) We claim that U_1 is the required neighbourhood. To see this let $g \in U_1$ and consider $\alpha(g)(\gamma)$ where $\gamma \in K$. Then $\gamma \in \gamma_i + P(\bar{U}, V_{\varepsilon/2})$, for some $i \in \{1, \ldots, m\}$. So $\gamma - \gamma_i \in P(\bar{U}, V_{\varepsilon/2})$. Thus $(\gamma - \gamma_i)(g) \in V_{\varepsilon/2}$ for $g \in U_1 \subseteq U$. As $\gamma_i(g) \in V_{\varepsilon/2}$, this implies that $\gamma(g) \in V_{\varepsilon/2} + V_{\varepsilon/2} \subseteq V_\varepsilon$. So $\alpha(g)(\gamma) \in V_\varepsilon$, as required. //

We continue the proof of the duality theorem in the next two chapters. In the first of these, the duality theorem is proved for compact groups and discrete groups. In the second it is extended to all LCA-groups.

There are a number of proofs of the duality theorem
in the literature. The most elegant appears in Rudin
(Fourier analysis on groups). Hewitt and Ross (Abstract
harmonic analysis) present the more classical approach of
first deriving the structure theory of LCA-groups and then
using it in the proof of duality. A fashionable proof is
given by D.W. Roeder, Category theory applied to Pontryagin
duality, *Pacific J.* 52 (1974), 519-527. Other references
include A. Weil (L'integration dans les groupes topologiques
et ses applications, *Actualités Sci. et Ind.*, Hermann &
Cie., 1951); H. Cartan and R. Godement, Théorie de la
dualité et analyse harmonique dans les groupes abéliens
localement compacta, *Ann. Sci. École Norm. Sup.* (3) 64
(1947), 79-99; D.A. Raikov, Harmonic analysis on commutative
groups with Haar measure and the theory of characters
(Russian), *Trudy Mat. Inst. Steklov* 14 (1945), German translation in Sowjetische Arbeiten zur Functionalanalysis 11-87,
Berlin: *Kultur u. Fortschritt*, 1954; M.A. Naimark (Normed
Rings, Nordhoff, 1959); and of course, L.S. Pontryagin
(Topological Groups).

EXERCISE SET EIGHT

1. Show that Z satisfies the duality theorem. (*Note*.
This requires more than just showing that Z^{**} is topologically isomorphic to Z. You must prove that the map α
in the duality theorem is a topological isomorphism. (Hint.
See Examples 1 and 2 in Chapter 3.)

2. Show that T satisfies the duality theorem. (Hint.
Firstly show that α is $1-1$ and onto. Then use the Open
Mapping Theorem.)

3. Prove that every discrete finite cyclic group satisfies the duality theorem.

4. Prove that R satisfies the duality theorem.

$$* * * * * * * * *$$

In the remainder of this chapter we make some observations which are needed in the proof of the duality theorem, but which are also of interest in themselves.

Theorem 13. *If $G_1,\ldots,G_n$ are LCA-groups with dual groups $\Gamma_1,\ldots,\Gamma_n$, respectively, then $\Gamma = \Gamma_1 \times \Gamma_2 \times \ldots \times \Gamma_n$ is the dual group of $G_1 \times G_2 \times \ldots \times G_n$.*

Proof. It suffices to prove this for the case $n = 2$. If $g = g_1 + g_2$ is the unique representation of $g \in G$ as a sum of elements of G_1 and G_2, then the pair $\gamma_1 \in \Gamma_1$ and $\gamma_2 \in \Gamma_2$ determine a character $\gamma \in \Gamma$ by the formula

(1) $\quad (g,\gamma) = (g_1,\gamma_1) + (g_2,\gamma_2)$

Since every $\gamma \in \Gamma$ is completely determined by its action on the subgroups G_1 and G_2, (1) shows that Γ is algebraically the direct sum of Γ_1 and Γ_2.

To see that Γ has the product topology $\Gamma_1 \times \Gamma_2$ simply note that

(a) $P(K,V_\varepsilon) \supseteq P(K_1,V_{\varepsilon/2}) + P(K_2,V_{\varepsilon/2})$, where K is any compact subset of $G = G_1 \times G_2$, $K_1 = p_1(K)$, $K_2 = p_2(K)$ and p_1 and p_2 are the projections of G onto G_1 and G_2, respectively,

and (b) if K_1 is a compact subset of G_1 containing 0 and K_2 is a compact subset of G_2 containing 0, then

56

$$P(K_1 \times K_2, V_\varepsilon) \subseteq P(K_1, V_\varepsilon) + P(K_2, V_\varepsilon) \ .$$

[See Problem 6 of Exercise Set Seven.] //

Corollary 1. *For each* $n \geqslant 1$, R^n *is topologically iso-morphic to its dual group.*

Corollary 2. *For each* $n \geqslant 1$, T^n *and* Z^n *are dual groups of each other.*

Corollary 3. *If* $G_1, \ldots, G_n$ *are LCA-groups which satisfy the duality theorem, then* $G_1 \times G_2 \times \ldots \times G_n$ *satisfies the duality theorem. Hence* $R^a \times T^b \times G$ *satisfies the duality theorem, where* G *is a discrete finitely generated abelian group, and* a *and* b *are non-negative integers.*

Proof. Exercise.

Theorem 13 shows that the dual group of a product is the product of the dual groups. We shall see, in due course, that the dual of a subgroup is a quotient group, and the dual of a quotient group is a subgroup. As a first step towards this we have Proposition 30.

Proposition 30. *Let* f *be a continuous homomorphism of an LCA-group* A *into an LCA-group* B . *Let a map* $f^* : B^* \to A^*$ *be defined by putting* $f^*(\gamma)(a) = \gamma f(a)$, *for each* $\gamma \in B^*$ *and* $a \in A$. *Then* f^* *is a continuous homomorphism of* B^* *into* A^* . *If* f *is onto then* f^* *is one-one. If* f *is both an open mapping and one-one then* f^* *is onto.*

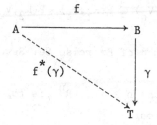

Proof. The verification that f^* is a homomorphism of B^* into A^* is routine. To see that f^* is continuous, let $P(K,U)$ be a subbasic open set in A^*, where U is an open subset of T and K is a compact subset of A. The continuity of f^* follows from the fact that $(f^*)^{-1}(P(K,U)) = P(f(K),U)$ is an open subset of B^*.

Assume f is onto and suppose that $f^*(\gamma_1) = f^*(\gamma_2)$, where γ_1 and γ_2 are in B^*. Then $f^*(\gamma_1(a)) = f^*(\gamma_2(a))$, for all $a \in A$; that is, $\gamma_1 f(a) = \gamma_2 f(a)$, for all $a \in A$. As f is onto this says that $\gamma_1(b) = \gamma_2(b)$, for all $b \in B$. Hence $\gamma_1 = \gamma_2$ and f^* is one-one.

Assume that f is both an open mapping and one-one. Let $\delta \in A^*$. As f is one-one, Proposition 17 tells us that there is a (not necessarily continuous) homomorphism $\gamma: B \to T$ such that $\delta = \gamma f$. As δ is continuous and f is an open mapping, γ is indeed continuous; that is, $\gamma \in B^*$. As $f^*(\gamma) = \delta$, we have that f^* is onto. //

Corollary 1. *If B is a quotient group of A, where $A and B$ are either both compact Hausdorff abelian groups or discrete abelian groups, then B^* is topologically isomorphic to a subgroup of A^*.*

Proof. Exercise.

Corollary 2. *If A is a subgroup of B, where A and B are discrete abelian groups, then A* is a quotient group* *of* B* .

Proof.   Exercise.

Remark.   As noted earlier we shall see in due course that
Corollary 1 and Corollary 2 remain true if the hypotheses
"compact Hausdorff" and "discrete" are replaced by "locally
compact Hausdorff".

The next lemma indicates that, before proving the duality
theorem, we shall have to see that LCA-groups have enough
characters to separate points.

Lemma.   In the notation of the duality theorem, the map α
is one-one if and only if G has enough characters to
separate points; that is, for each g and h in G , with
g ≠ h , there is a γ ∈ Γ such that γ(g) ≠ γ(h) .

Proof.   Assume that α is one-one. Suppose that there
exist g and h in G , with g ≠ h , such that γ(g) = γ(h)
for all γ ∈ Γ . Then α(g)(γ) = α(h)(γ) , for all γ ∈ Γ .
So α(g) = α(h) , which implies that g = h , a contradiction.
Hence G has enough characters to separate points.
   Assume now that G has enough characters to separate
points. Let g and h be in G , with g ≠ h . Then there
is a γ ∈ Γ such that γ(g) ≠ γ(h) . So α(g)(γ) ≠ α(h)(γ) ,
which implies that α(g) ≠ α(h) . So α is one-one. //

The final proposition in this chapter should remind the
reader of the Stone-Weierstrass Theorem.

Proposition 31.   *Let* G *be an LCA-group and* Γ *its dual*
group. *Let* G *satisfy the duality theorem and also have*

the property that every non-trivial Hausdorff quotient group Γ/B of Γ has a non-trivial character. If A is a subgroup of Γ which separates points of G then A is dense in Γ .

Proof. Suppose A is not dense in Γ . If B is the closure of A in Γ then Γ/B is a non-trivial LCA-group. So there exists a non-trivial continuous homomorphism φ: Γ/B → T . Let f be the canonical homomorphism : Γ → Γ/B . Then φf is a continuous homomorphism : Γ → T . Furthermore, φf(Γ) ≠ 0 but φf(B) = 0 . As G satisfies the duality theorem, there is a g ∈ G such that φf(γ) = γ(g) , for all γ ∈ Γ . So γ(g) = 0 for all γ in A . But since A separates points in G , this implies g = 0 . So φf(Γ) = 0 , which is a contradiction. Hence A is dense in Γ . //

Of course the second sentence in the statement of Proposition 31 will, in due course, be seen to be redundant.

EXERCISE SET NINE
1. (i)   Show that if $G_1, \ldots, G_n$ are LCA-groups which satisfy the duality theorem, then $G_1 \times G_2 \times \ldots \times G_n$ satisfies the duality theorem.
    (ii) Deduce that every discrete finitely generated abelian group satisfies the duality theorem. (Hint: Use the fact that every finitely generated abelian group is a direct product of a finite number of cyclic groups.)
    (iii) Hence show that $R^a \times T^b \times G$ satisfies the duality theorem, where G is a discrete finitely generated abelian group, and a and b are non-negative integers.

2. Show that if B is a quotient group of A, where A and B are either both compact Hausdorff abelian groups or discrete abelian groups, then B^* is topologically isomorphic to a subgroup of A^*.

3. Show that if A is a subgroup of B, where A and B are discrete abelian groups, then A^* is a quotient group of B^*.

4. Show that if G is any LCA-group and Γ is its dual group, then Γ has enough characters to separate points.

* * * * * * * * *

5 · Duality for compact and discrete groups

In the last chapter (see the Lemma preceding Proposition 31) we saw that a necessary condition for a topological group to satisfy duality is that it have enough characters to separate points. That discrete abelian groups have this property has been indicated already in the Corollary of Proposition 17. For compact groups we must borrow a result from the representation theory of topological groups. [For a brief outline of this theory, see Higgins (An Introduction to Topological Groups). Fuller discussions appear in Adams (Lectures on Lie Groups), Hewitt and Ross (Abstract Harmonic Analysis I), Pontryagin (Topological Groups) and Hotmann, *Proc.Camb.Philos.Soc.* 65 (1969) 47-52.]

Theorem (Peter, Weyl, van Kampen) *Let* G *be a compact Hausdorff group. Then* G *has sufficiently many irreducible continuous representations by unitary matrices. In other words, for each* $g \in G$, $g \neq e$, *there is a continuous homomorphism* ϕ *of* G *into the unitary group* U(n) , *for some* n , *such that* $\phi(g) \neq e$.

If G is abelian then, without loss of generality, it can be assumed that n = 1 . As U(1) = T we obtain the following theorem.

Theorem 14. *Every compact Hausdorff abelian topological group has enough characters to separate points.*

Theorem 14 was first proved by John von Neumann for compact metrizable abelian groups. A derivation of Theorem 14 from von Neumann's result is outlined in Problems 2 and 3 of Exercise Set Ten.

Corollary 1. *Let* G *be any compact Hausdorff abelian group. Then* G *is topologically isomorphic to a closed subgroup of the product* $\prod_{i \in I} T_i$, *where each* T_i *is topologically isomorphic to* T , *and* I *is some index set.*

Proof. Exercise.

Corollary 2. *Let* G *be a compact Hausdorff abelian group. Then every neighbourhood* U *of* 0 *contains a closed subgroup* H *such that* G/H *is topologically isomorphic to* $T^n \times D$, *for some finite discrete group* D *and* $n \geqslant 0$.

Proof. Exercise.

EXERCISE SET TEN

1. Using Problem 4(iii) of Exercise Set Three, show that every compact totally disconnected abelian topological group has enough characters to separate points.

2. Show that every compactly generated locally compact Hausdorff group G can be approximated by metrizable groups in the following sense: For each neighbourhood U of e , there exists a compact normal subgroup H of G such that H ⊆ U and G/H is metrizable.
(Hint: Let $V_1, V_2, V_3, \ldots$ be a sequence of symmetric compact neighbourhoods of e such that (i) $V_1 \subseteq U$, (ii) $V_{n+1}^2 \subseteq V_n$, for $n \geqslant 1$, and (iii) $g^{-1} V_n g \subseteq V_{n-1}$, for $n \geqslant 2$ and $g \in K$, where K is a compact set which generates G . Put $H = \bigcap_{n=1}^{\infty} V_n$ and use Problem 3(ii) of Exercise Set Six.)

3. Using Problem 2 above, deduce statement B from

statement A.

    (A)   Every compact metrizable abelian group has enough characters to separate points.

    (B)   Every compact Hausdorff abelian group has enough characters to separate points.

4. (i)   Show that every compact Hausdorff abelian group G is topologically isomorphic to a subgroup of a product $\prod_{i \in I} T_i$ of copies of T.
(Hint: See the proof of Theorem 1.)

  (ii) If G is also metrizable show that the index set I can be chosen to be a countable set.
(Hint: Use Problem 3(ii) of Exercise Set Six.)

  (iii) Using (i) show that if G is any compact Hausdorff abelian group, then every neighbourhood U of O contains a closed subgroup H such that G/H is topologically isomorphic to $T^n \times D$, for some finite discrete group D and $n \geqslant 0$.
(Hint: Reread "Remarks on products" in Chapter 1. Use Corollary 3 of Theorem 7.)

5. Show that every compact Hausdorff group is topologically isomorphic to a subgroup of a product of copies of U, where $U = \prod_{n=1}^{\infty} U(n)$.

$$* \quad * \quad * \quad * \quad * \quad * \quad * \quad * \quad *$$

The next proposition provides the last piece of information we need in order to prove the duality theorem for compact groups and discrete groups. (This proposition should be compared with Proposition 31.)

Proposition 32. *Let* G *be a discrete abelian group and* Γ *its dual group. If* A *is a subgroup of* Γ *which separates points of* G *, then* A *is dense in* Γ *.*

Proof. Noting how the topology on Γ is defined, it suffices to show that each non-empty sub-basic open set $P(K,U)$, where K is a compact subset of G and U is an open subset of T , intersects A non-trivially.

As G is discrete, K is finite. Let H be the sub-group of G generated by K and $f^*: \Gamma \rightarrow H^*$ the map obtained by restricting the characters of G to H . According to Proposition 30 and its Corollary 2, f^* is an open continuous homomorphism of Γ onto H^* . As A separates points of G , $f^*(A)$ separates points of H . Observing that Corollary 3 of Theorem 13 says that H satisfies the duality theorem, Proposition 31 then implies that $f^*(A)$ is dense in H^* . So $f^*(P(K,U)) \cap f^*(A) \neq \phi$. In other words there is a $\gamma \in A$ such that, when restricted to H , γ maps K into U . Of course this says that $\gamma \in P(K,U) \cap A$. //

Theorem 15. *Let* G *be a compact Hausdorff abelian group and* Γ *its dual group. Then the canonical map* α *of* G *into* Γ^* *is a topological group isomorphism of* G *onto* Γ^* *.*

Proof. By the two Lemmas in Chapter 4 together with Theorem 14, α is a continuous one-one homomorphism of G into Γ^* . Clearly $\alpha(G)$ separates points of Γ . As Γ is discrete, Proposition 32 then implies that $\alpha(G)$ is dense in Γ^* . However $\alpha(G)$ is compact and hence closed in Γ^* . Thus $\alpha(G) = \Gamma^*$; that is, the map α is onto. Finally, the Open Mapping Theorem tells us that α is also an open map. //

Corollary 1. *Let* G *be a compact Hausdorff abelian group and* Γ *its dual group. If* A *is a subgroup of* Γ *which separates points of* G *, then* A = Γ *.*

Proof. This is an immediate consequence of Theorem 15, Proposition 31, the Corollary of Proposition 17, and Theorem 12. //

Corollary 2. *Let* G *be an LCA-group with enough characters to separate points and* K *a compact subgroup of* G *. Then every character of* K *extends to a character of* G *.*

Proof. The collection of characters of K which extend to characters of G form a subgroup A of K^* . As G has enough characters to separate points, A separates points of K . So by Corollary 1 above, $A = K^*$. //

Corollary 3. *Let* B *be an LCA-group with enough characters to separate points and* f *a continuous one-one homomorphism of a compact group* A *into* B *. Then the map* $f^*: B^* \to A^*$ *, described in Proposition 30, is a quotient homomorphism.*

Proof. Exercise.

Theorem 16. *Let* G *be a discrete abelian group and* Γ *its dual group. Then the canonical map* α *is a topological group isomorphism of* G *onto* $Γ^*$ *.*

Proof. As in Theorem 15, α is a continuous one-one homomorphism of G into $Γ^*$. As α(G) separates points of Γ and Γ is compact, Corollary 1 above yields that α(G) = Γ . As G and $Γ^*$ are discrete this completes the proof. //

We conclude this chapter by showing how duality theory
yields a complete description of compact Hausdorff abelian
torsion groups. (Recall that a group G is said to be a
torsion group if each of its elements is of finite order.)
The first step is the following interesting result.

Theorem 17. *If G is the direct product of any family*
{G_i: i $\in$ I} of compact Hausdorff abelian groups, then the
discrete group G^* is algebraically isomorphic to the
restricted directed product of the corresponding dual groups
{Γ_i: i $\in$ I} .

Proof. Each $g \in G$ may be thought of as a "string" g =
$(\ldots, g_i, \ldots)$, the group operation being componentwise
addition. If $\gamma = (\ldots, \gamma_i, \ldots)$, where $\gamma_i \in \Gamma_i$ and only
finitely many γ_i are non-zero, then γ is a character on
G defined by $(g, \gamma) = \sum_{i \in I} (g_i, \gamma_i)$, for each $g \in G$.
(Observe that this is a finite sum!) Let us denote the
subgroup of G^* consisting of all such γ by A . Then
A is algebraically isomorphic to the restricted direct
product of {Γ_i: i $\in$ I} .

We claim that A separates points of G . To see this
let $g \in G$, $g \neq 0$. Then $g = (\ldots, g_i, \ldots)$ with some
$g_i \neq 0$. So there is a $\gamma_i \in \Gamma_i$ such that $\gamma_i(g_i) \neq 0$.
Putting $\gamma = (\ldots, \gamma_j, \ldots)$ where $\gamma_j = 0$ unless j = i ,
we see that $\gamma(g) = \gamma_i(g_i) \neq 0$. As $\gamma \in A$, A separates
points of G . By Corollary 1 of Theorem 15, this implies
that $A = G^*$. //

Corollary. *Every countable abelian group is algebraically*
isomorphic to a quotient group of a countable restricted
direct product of copies of Z .

Proof. Let G be a countable abelian group. Put the

discrete topology on G and let Γ be its dual group. Of course Γ is compact and by Problem 4(i) of Exercise Set Ten, Γ is topologically isomorphic to a subgroup of a product $\prod_{i\in I} T_i$ of copies of T , where the cardinality of the index set I equals the cardinality of Γ^* . By Theorem 16, Γ^* is topologically isomorphic to G . So Γ is topologically isomorphic to a subgroup of a countable product of copies of T . Taking dual groups and using Theorem 17 and Corollary 3 of Theorem 15 we obtain the required result. //

Of course the above corollary also follows from the fact that the free abelian group on a countable set is a countable restricted direct product of copies of Z , and that any countable abelian group is a quotient group of the free abelian group on a countable set.

Remark.   S. Kaplan (Extensions of Pontryagin duality I and II, *Duke Math. J.* 15 (1948), 649-658 and 17 (1950), 419-435) has investigated generalizations of Theorem 17 to direct products of non-compact groups. As a direct product of LCA-groups is not, in general, an LCA-group we must first say what we mean by the dual group of a non-LCA-group: If G is any abelian topological group we define Γ to be the group of continuous homomorphisms of G into T , with the compact open topology. Then Γ is an abelian topological group and we can form Γ^* in the same way. As in the locally compact case there is a natural map α which takes $g \in G$ to α(g) a function from Γ into T . We can then ask for which groups is α a topological group isomorphism of G onto Γ^* . Such groups will be called *reflexive*. A satisfactory description of this class is not known, but it includes not only all LCA-groups but also all Banach spaces (considered as topological groups). (See M.F. Smith, The Pontryagin duality theorem in linear spaces, *Ann. of Math.*

(2) 56 (1952), 248-253.) Kaplan showed that if $\{G_i : i \in I\}$ is a family of reflexive groups then $\prod\limits_{i \in I} G_i$ is also a reflexive group. Its dual group is algebraically isomorphic to the restricted direct product of the family $\{G_i^* : i \in I\}$. The topology of the dual group is slightly complicated to describe, but when I is countable it is simply the subspace topology induced on $\prod\limits_{i \in I}^{r} G_i^*$ if $\prod\limits_{i \in I} G_i^*$ is given the box topology. In particular this is the case when each G_i is an LCA-group - thus generalizing Theorem 17.

For further comments on reflexive groups see R. Brown, P.J. Higgins and S.A. Morris, Countable products and sums of lines and circles; their closed subgroups and duality properties, *Math. Proc. Camb. Philos. Soc.* 78 (1975), 19-32; R. Venkataraman, Extensions of Pontryagin duality, *Math. Z.* 143 (1975), 105-112; N. Noble, k-groups and duality, *Trans. Amer. Math. Soc.* 151 (1970), 551-561; N.Th. Varopoulos, Studies in harmonic analysis, *Proc. Camb. Philos. Soc.* 60 (1964), 465-516; N.Ya. Vilenkin, The theory of characters of topological abelian groups with boundedness given, *Izv. Akad. Nauk. SSSR. Ser. Mat.* 15 (1951), 439-462.

To prove the structure theorem of compact Hausdorff abelian torsion groups we have to borrow the following result of abelian group theory. (See L. Fuchs, *Abelian groups*, Pergamon Press, 1960.)

Theorem. *An abelian group all of whose elements are of bounded order is algebraically isomorphic to a restricted direct product* $\prod\limits_{i \in I}^{r} Z(b_i)$, *with only a finite number of the* b_i *distinct where* $Z(b_i)$ *is the discrete cyclic group with* b_i *elements.*

Theorem 18. *A compact Hausdorff abelian torsion group is topologically isomorphic to* $\prod\limits_{i \in I} Z(b_i)$, *where* I *is some index set and there exist only a finite number of distinct* b_i .

Proof.   Exercise.

EXERCISE SET ELEVEN

1. If f is a continuous one-one homomorphism of a
compact group A into an LCA-group B which has enough
characters to separate points, show that the map $f^*: B^* \to A^*$,
described in Proposition 30, is a quotient homomorphism.

2. Show that every compact Hausdorff abelian torsion
group G is topologically isomorphic to a product
$\prod_{i \in I} Z(b_i)$, where $Z(b_i)$ is a discrete cyclic group with
b_i elements, I is an index set, and where there are only
a finite number of distinct b_i .
(Hint: Let $G_{(n)} = \{x \in G : nx = 0\}$. Observe that
$G = \bigcup_{n=1}^{\infty} G_{(n)}$ and using the Baire-Category Theorem show that
one of the quotient groups $G/G_{(n)}$ is finite. Deduce that
the orders of all elements of G are bounded. Then use the
structure theorem of abelian groups of bounded order.)

* * * * * * * * *

6 · The duality theorem and the principal structure theorem

In Chapter 5 we proved the duality theorem for compact groups and discrete groups. To extend the duality theorem to all LCA-groups we will prove two special cases of the following proposition: If G is an LCA-group with a subgroup H such that both H and G/H satisfy the duality theorem, then G satisfies the duality theorem. The two cases we prove are when H is compact and when H is open. The duality theorem for all LCA-groups then follows from the fact that every LCA-group G has an open subgroup H which in turn has a compact subgroup K such that H/K is an "elementary group" which is known to satisfy the duality theorem. By an "elementary group" we mean one which is of the form $R^a \times Z^b \times T^c \times F$, where F is a finite discrete abelian group and a, b and c are non-negative integers. Once we have the duality theorem we use it, together with the above structural result, to prove the Principal Structure Theorem.

We begin with some structure theory.

Definition. A topological group is said to be *monothetic* if it has a dense cyclic subgroup.

Examples. Z and T are monothetic.

Theorem 19. *Let G be a monothetic LCA-group. Then either G is compact or G is topologically isomorphic to Z.*

Proof. If G is discrete then either $G = Z$ or G is a finite cyclic group and hence is compact. So we have to prove that G is compact if it is not discrete.

Assume G is not discrete. Then the dense cyclic sub-group $\{x_n : n = 0, \pm 1, \pm 2, \ldots\}$, where $x_n + x_m = x_{n+m}$ for each n and m, is infinite. (If the cyclic subgroup were finite it would be discrete and hence closed in G. As it is also dense in G, this would mean that it would equal G and G would be discrete.)

Let V be an open symmetric neighbourhood of 0 in G with $\bar{V}$ compact. If $g \in G$ then $V + g$ contains some x_k. Then there is a symmetric neighbourhood W of 0 in G such that $(g - x_k) + W \subseteq V$. As G is not discrete, W contains an infinite number of the x_n's and as W is symmetric $x_{-n} \in W$ if $x_n \in W$. Hence there exists a $j < k$ such that $x_j \in W$. Putting $i = k - j$ we have $i > 0$ and

$$g - x_i = g - x_k + x_j \in g - x_k + W \subseteq V.$$

This proves that $G = \bigcup\limits_{i=1}^{\infty} (x_i + V)$. (The important point is that we only need x_i, $\underline{i > 0}$.) As $\bar{V}$ is a compact subset of G we have that

$$(1) \quad \bar{V} \subseteq \bigcup\limits_{i=1}^{N} (x_i + V) \quad , \quad \text{for some } N .$$

For each $g \in G$, let $n = n(g)$ be the smallest positive n such that $g \in x_n + \bar{V}$. By (1), $x_n - g \in x_i + \bar{V}$ for some $1 \leq i \leq N$, so we have that $g \in x_{n-i} + \bar{V}$. Since $i > 0$, $n - i < n$ and so by our choice of n, $n - i \leq 0$. Thus $n \leq i \leq N$. So for each $g \in G$, $n \leq N$, which means that

$$G = \bigcup\limits_{i=1}^{N} (x_i + \bar{V})$$

which is a finite union of compact sets and so G is compact. //

Theorem 20. *A compact Hausdorff abelian group* G *is mono-thetic if and only if* G^* *is topologically isomorphic to a subgroup of* T_d *, the circle group endowed with the discrete topology.*

Proof. Exercise.

We now use Theorem 19 to obtain our first description of the structure of compactly generated LCA-groups. (Recall that an LCA-group G is said to be *compactly generated* if it has a compact subset V such that G is generated algebraically by V . Without loss of generality V can be chosen to be a symmetric neighbourhood of 0 .)

Proposition 33. *If* G *is an LCA-group which is algebraically generated by a compact symmetric neighbourhood* V *of* 0 *, then* G *has a closed subgroup* A *topologically isomorphic to* Z^n *, for some* $n \geqslant 0$ *, such that* G/A *is compact and* $V \cap A = \{0\}$ *.*

Proof. If we put $V_1 = V$ and $V_{n+1} = V_n + V$, for each $n \geqslant 1$, then $G = \bigcup_{n=1}^{\infty} V_n$.

As V_2 is compact there are elements $g_1, \ldots, g_m$ in G such that $V_2 \subseteq \bigcup_{i=1}^{m} (g_i + V)$. Let H be the group generated by $\{g_1, \ldots, g_m\}$. So $V_i \subseteq V + H$, for $i = 1$ and $i = 2$. If we assume that $V_n \subseteq V + H$, then we have

$$V_{n+1} \subseteq V + (V + H) = V_2 + H \subseteq (V + H) + H = V + H .$$

So, by induction, $V_n \subseteq V + H$, for all $n \geqslant 1$, and hence $G = V + H$.

Let $\bar{H}_i$ be the closure in G of the subgroup H_i generated by g_i , for $i = 1, \ldots, m$. As $H = H_1 + \ldots + H_m$, if each $\bar{H}_i$ is compact, then $\bar{H}$ is compact and so $G = V + \bar{H}$

is compact. (Use Exercise Set One, Problem 4 (iii).) The
Proposition would then be true with n = 0 . If G is not
compact, then, by Theorem 19, one of the monothetic groups
$\bar{H}_i$ is topologically isomorphic to Z . In this case
$\bar{H}_i = H_i$ and we deduce that

$$*\begin{cases} \text{If } G = V + H \text{, where } H \text{ is a finitely generated} \\ \text{group, and } G \text{ is not compact, then } H \text{ has a sub-} \\ \text{group topologically isomorphic to } Z \text{.} \end{cases}$$

As H is a finitely generated abelian group (and every
subgroup of an abelian group with p generators can be
generated by $\leq$ p elements) there is a largest n such
that H contains a subgroup A topologically isomorphic
to Z^n . Since A is discrete and V is compact, $A \cap V$
is finite. Without loss of generality we can assume that
$A \cap V = \{0\}$. (If necessary we replace A by a subgroup
A' which is also topologically isomorphic to Z^n and has
the property that $A' \cap V = \{0\}$. For example, if A =
$gp\{a_1,\ldots,a_n\}$ and r is chosen such that $A \cap V \subseteq \{k_1 a_1 +$
$\ldots + k_n a_n : 1 - r \leq k_i \leq r - 1 , i = 1,\ldots,n\}$ then we put
$A' = gp\{ra_1,\ldots,ra_n\}$.)
   Let f be the canonical homomorphism of G onto K = G/A .
Then K = f(V) + f(H) . By Problem 2 of Exercise Set Twelve
and our choice of n , f(H) has no subgroup topologically
isomorphic to Z . By (*) applied to K instead of G , we
see that K is compact, as required. //

   The above proposition allows us to prove a most important
theorem which, of course, generalizes Theorem 14 and the
Corollary to Proposition 17.

Theorem 21. *Every LCA-group has enough characters to
separate points.*

74

Proof. Let G be any LCA-group and g any non-zero element of G . Let V be a compact symmetric neighbourhood of O which contains g . Then the subgroup H generated algebraically by V is, by Proposition 8, an open subgroup of G . By Proposition 33, H has a closed subgroup A such that H/A is compact and V ∩ A = {0} . Defining f to be the canonical map of H onto H/A we see that f(g) ≠ O .

According to Theorem 14 there is a continuous homomorphism φ: H/A → T such that φ(f(g)) ≠ O . Then φf is a continuous homomorphism of H into T . As H is an open subgroup of G and T is divisible, Proposition 17 tells us that φf can be extended to a continuous homomorphism γ: G → T . Clearly γ(g) ≠ O and so G has enough characters to separate points. //

Corollary 1. *Let* H *be a closed subgroup of an LCA-group* G . *If* g *is any element of* G *not in* H , *then there is a character* γ *of* G *such that* γ(g) ≠ O *but* γ(h) = O , *for all* h ∈ H .

Proof. Exercise.

The next corollary is an immediate consequence of the opening sentences in the proof of Theorem 21.

Corollary 2. *Every LCA-group has a subgroup which is both open and a compactly generated LCA-group.*

Remarks. Theorem 21 was first proved by E.R. van Kampen. A proof based on the theory of Banach algebras was given by I.M. Gelfand and D.A. Raikov.

The reader should not be misled, by Theorem 21, into thinking that all Hausdorff abelian topological groups have enough characters to separate points. This is not so. See Section 23.32 of E. Hewitt and K.A. Ross, Abstract Harmonic Analysis I.

The next proposition gives another useful description of the structure of compactly generated LCA-groups.

Proposition 34. *If* G *is a compactly generated LCA-group, then it has a compact subgroup* K *such that* G/K *is topologically isomorphic to* $R^a \times Z^b \times T^c \times F$ *, where* F *is a finite discrete abelian group and* a *,* b *and* c *are nonnegative integers.*

Proof. By Proposition 33 there exists a discrete finitely generated subgroup D of G such that G/D is compact. Let N be a compact symmetric neighbourhood of 0 such that $3N \cap D = \{0\}$. If f: $G \to G/D$ is the canonical homomorphism then f(N) is a neighbourhood of 0 in G/D and, by Corollary 2 of Theorem 14, there exists a closed subgroup $B \subseteq f(N)$ such that (G/D)/B is topologically isomorphic to $T^n \times E$, where E is a finite discrete group and $n \geqslant 0$. If we let $K' = f^{-1}(B)$ then we see that G/K' is topologically isomorphic to $T^n \times E$.

Putting $K = K' \cap N$, we have that K is compact and f(K) = B . To see that K is a subgroup of G , let x and y be in K . Then $x - y \in K'$, so there is a $z \in K$ such that f(z) = f(x - y) . This implies that $x - y - z \in D$, and since $3N \cap D = \{0\}$ it follows that x - y - z = 0 ; that is, $x - y \in K$ and so K is a subgroup of G . We claim that K' = K + D . For if $k' \in K'$, there is a $k \in K$ such that f(k') = f(k) and so $k' - k \in D$. Thus K' = K + D . By Problem 8 of Exercise Set Four, K' is topologically isomorphic to $K \times D$. Hence if θ is the canonical map of G onto G/K then $\theta(D)$ is topologically isomorphic to D and (G/K) / $\theta(D)$ is topologically isomorphic to G/K' which

is in turn topologically isomorphic to $T^n \times E$. As $\theta(D)$
and E are discrete, Problem 6 of Exercise Set Five tells
us that G/K is locally isomorphic to T^n and hence also
to R^n . Theorem 8 then says that G/K is topologically
isomorphic to $R^a \times T^c \times S$, where S is a discrete group
and $a \geqslant 0$ and $c \geqslant 0$. As G is compactly generated
G/K and hence also S are compactly generated. So S is
a discrete finitely generated abelian group and thus is
topologically isomorphic to $Z^b \times F$, for some finite
discrete group F and $b \geqslant 0$. //

EXERCISE SET TWELVE

1. (i)   Let f be a continuous homomorphism of an
         LCA-group A into an LCA-group B . If $f(A)$
         is dense in B , show that the map $f^* : B^* \to A^*$,
         described in Proposition 30, is one-one.

   (ii) Show that if G is a compact Hausdorff abelian
         group which is monothetic then G^* is topo-
         logically isomorphic to a subgroup of T_d , the
         circle group endowed with the discrete topology.
         (Hint: Use (i) with $A = Z$ and $B = G$.)

   (iii) Let A be an LCA-group which satisfies the
         duality theorem and B an LCA-group. If f
         is a continuous one-one homomorphism of A into
         B show that $f^*(B^*)$ is dense in A^* .
         (Hint: See the proof of Corollary 2 of Theorem
         15 and use Proposition 31 and Theorem 21.)

   (iv) Show that if G is a compact Hausdorff abelian
         group with G^* topologically isomorphic to a
         subgroup of T_d , then G is monothetic.

2. Let A and B be LCA-groups and H a (not necessarily
closed) finitely generated subgroup of A . If f is a con-

tinuous homomorphism of A into B such that the kernel
of f lies wholly in H and is topologically isomorphic
to Z^n , for some $n \geqslant 1$, and such that $f(H)$ contains a
subgroup topologically isomorphic to Z , show that H
contains a subgroup topologically isomorphic to Z^{n+1} .
(Hint: Use the Corollary of Theorem 2.)

3. If H is a closed subgroup of an LCA-group G and
g is an element of G not in H , show that there is a
character γ of G such that $\gamma(g) \neq 0$ but $\gamma(h) = 0$,
for all $h \in H$.

4. A Hausdorff topological space X is said to be a
k_ω-space if $X = \bigcup\limits_{n=1}^{\infty} X_n$, where (a) each X_n is compact;
(b) $X_n \subseteq X_{n+1}$, for each n ; (c) a subset A of X is
closed in X if and only if $A \cap X_n$ is compact for each
n . Prove

   (i)   R is a k_ω-space.
   (ii)   A locally compact Hausdorff group G is a k_ω-
          space if and only if it is σ-compact. If G
          is σ-compact, show that the X_n's in the k_ω-
          decomposition can be chosen to be neighbour-
          hoods of e .
   (iii) Any connected locally compact Hausdorff group
          is a k_ω-space.
   (iv)   If $X = \bigcup\limits_n X_n$ is a k_ω-space, then any compact
          subset K of X is contained in some X_n .
(Hint for (ii): As G is σ-compact, $G = \bigcup\limits_{n=1}^{\infty} Y_n$ where each
Y_n is compact. Let V be a compact symmetric neighbourhood
of e and put $X_n = Y_1 V \cup Y_2 V \ldots \cup Y_n V$.)

5. (i)   Let G be a locally compact Hausdorff group
          and N a closed normal subgroup of G . If
          f: G → G/N is the canonical map, show that for

each compact subset C of G/N there exists a compact subset S of G such that $f(S) = C$.

(ii) Deduce that if N is a closed normal subgroup of a locally compact Hausdorff group G such that both N and G/N are compactly generated, then G is also compactly generated.

(iii) If in (i), N is also compact show that $f^{-1}(C)$ is compact.

(iv) Deduce that if G is a Hausdorff topological group having a normal subgroup K such that both K and G/K are compact, then G is compact.

* * * * * * * * *

Definition. Let A , B and C be topological groups, f_1 a continuous homomorphism of A into B and f_2 a continuous homomorphism of B into C . The sequence

$$0 \longrightarrow A \xrightarrow{\ f_1\ } B \xrightarrow{\ f_2\ } C \longrightarrow 0$$

is said to be *exact* if

(i) f_1 is one-one,

(ii) f_2 is onto,

and (iii) the kernel of f_2 equals $f_1(A)$.

Proposition 35. *Let K be a compact subgroup of an LCA-group G , so that we have an exact sequence*

$$0 \longrightarrow K \xrightarrow{\ f_1\ } G \xrightarrow{\ f_2\ } G/K \longrightarrow 0$$

where f_2 is an open continuous homomorphism and f_1 is a *homeomorphism of* K *onto its image in* G . *Then the sequence*

$$0 \longleftarrow K \overset{f_1^*}{\longleftarrow} G \overset{f_2^*}{\longleftarrow} (G/K)^* \longleftarrow 0$$

is exact and f_1^* *and* f_2^* *are open continuous homomorphisms.*

Proof. By Proposition 30, f_2^* is one-one. Using Corollary 3 of Theorem 15 together with Theorem 21 we see that f_1^* is both open and onto. To see that the image of f_2^* equals the kernel of f_1^* consider the diagram

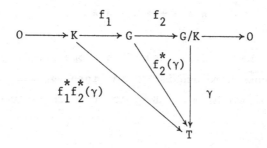

Let γ be any character of G/K and k any element of K. Then $f_1^* f_2^* \gamma(k) = \gamma f_2 f_1(k) = 0$ as the given sequence is exact. Therefore $f_1^* f_2^*(\gamma) = 0$ and so Image $f_2^* \subseteq$ Kernel f_1^*. Now if $\phi \in G^*$ and $f_1^*(\phi) = 0$, then we have $\phi f_1(k) = 0$ for all $k \in K$. So there exists a homomorphism $\delta: G/K \to T$ such that $\delta f_2 = \phi$. As f_2 is both open and onto, δ is continuous. So Kernel $f_1^* \subseteq$ Image f_2^*. Hence Image $f_2^* =$ Kernel f_1^*.

Finally we have to show that f_2^* is an open map. Let C be a compact subset of G/K, U an open subset of T and P(C,U) the set of all elements of $(G/K)^*$ which map C into U. Then P(C,U) is a sub-basic open set in $(G/K)^*$. Now by Problem 5 (i) of Exercise Set Twelve there exists a compact subset S of G such that $f_2(S) = C$. Thus we see that $P(S,U)$ is a sub-basic open subset of G^* such that $f_2^*(P(C,U)) = P(S,U) \cap f_2^*((G/K)^*)$. So f_2^* is a homeomorphism of $(G/K)^*$ onto its image in G^* . As K^* is discrete, Kernel f_1^* is open in G^* ; that is, Image f_2^* is open in G^* . So f_2^* is an open map. //

Proposition 36. *Let* A *be an open subgroup of an LCA-group* G , *so that we have an exact sequence*

$$0 \xrightarrow{} A \xrightarrow{f_1} G \xrightarrow{f_2} G/A \xrightarrow{} 0$$

where the homomorphisms f_1 *and* f_2 *are open continuous maps. Then the sequence*

$$0 \xleftarrow{} A^* \xleftarrow{f_1^*} G^* \xleftarrow{f_2^*} (G/A)^* \xleftarrow{} 0$$

is exact, f_1^* *is open and continuous and* f_2^* *is a homeomorphism of* $(G/A)^*$ *onto its image in* G^* .

Proof. By Proposition 30, f_1^* is onto and f_2^* is one-one. That Image f_2^* = Kernel f_1^* is proved exactly as in Proposition 35. As A is open in G , G/A is discrete and $(G/A)^*$ is compact. As f_2^* is one-one and $(G/A)^*$ is compact, f_2^* is a homeomorphism of $(G/A)^*$ onto its image in G^* .

Finally we have to show that f_1^* is an open map. Let K be a compact neighbourhood of 0 in G which lies in A . If V_a is as in Theorem 11, then $P(K,V_a)$ is an open set in G^* such that $\overline{P(K,V_a)}$ is compact. Of course, $f_1^*(P(K,V_a))$ consists of those elements of A which map K into V_a , and so is open in A^* . If we put H equal to the group generated by $f_1^*(P(K,V_a))$ then H is an open subgroup of A^*. Furthermore as $gp\{P(K,V_a)\}$ is an open and closed subgroup of G^*, $\overline{P(K,V_a)} \subseteq gp\{P(K,V_a)\} = B$. As B is generated by $\overline{P(K,V_a)}$ it is σ-compact. The Open Mapping Theorem then implies that $f_1^*: B \to H$ is open. As B is an open subgroup of G^* and H is an open subgroup of A^*, $f_1^*: G^* \to A^*$ is open. //

The next Proposition is a corollary of the 5-Lemma of category theory. It is easily verified by "diagram-chasing".

Proposition 37. *Let* A , B , C , D , E *and* F *be abelian topological groups and* f_1 , f_2 , f_3 , f_4 , f_5 , f_6 *and* f_7 *be continuous homomorphisms as indicated in the diagram below.*

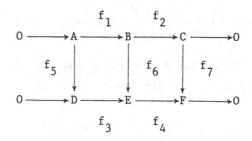

Let each of the horizontal sequences be exact and let the diagram be commutative (that is, $f_3 f_5 = f_6 f_1$ *and* $f_4 f_6 = f_7 f_2$ *). If* f_5 *and* f_7 *are algebraic isomorphisms (that is, both one-one and onto) then* f_6 *is also an algebraic isomorphism.*

We now prove the duality theorem for compactly generated LCA-groups.

Theorem 22. *Let* G *be a compactly-generated LCA-group and*

Γ *its dual group. Then the canonical map* α *of* G *into* $\Gamma^*{}^*$ *is a topological group isomorphism of* G *onto* $\Gamma^*{}^*$.

Proof. By Proposition 34, G has a compact subgroup K such that G/K is topologically isomorphic to $R^a \times Z^b \times T^c \times F$ where F is a finite discrete abelian group and a , b and c are non-negative integers. So we have an exact sequence

$$0 \xrightarrow{} K \xrightarrow{f_1} G \xrightarrow{f_2} G/K \xrightarrow{} 0$$

Applying Proposition 35 to this sequence and Proposition 36 to the dual sequence, we obtain that the sequence

$$0 \xrightarrow{} K^{**} \xrightarrow{f_1^{**}} \Gamma^* \xrightarrow{f_2^{**}} (G/K)^{**} \xrightarrow{} 0$$

is also exact. It is easily verified that the diagram

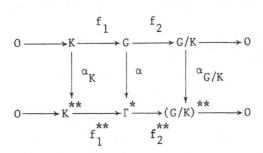

is commutative, where α_K and $\alpha_{G/K}$ are the canonical maps. As we have already seen that K and G/K satisfy the duality theorem, α_K and $\alpha_{G/K}$ are both topological isomorphisms. This implies, by Proposition 37, that α is an algebraic isomorphism. As α is continuous and G is compactly generated the Open Mapping Theorem then implies that α is an open map, and hence a topological isomorphism. //

We now prove the duality theorem for all LCA-groups.

Theorem 23 (Pontryagin-van Kampen Duality Theorem). *Let G be an LCA-group and Γ its dual group. Then the canonical map α of G into Γ^{*} is a topological group isomorphism of G onto Γ^{*}.*

Proof. By Corollary 2 of Theorem 21, G has an open subgroup A which is compactly generated. So we have an exact sequence

$$0 \longrightarrow A \xrightarrow{\;f_1\;} G \xrightarrow{\;f_2\;} G/A \longrightarrow 0$$

Applying Proposition 36 and then Proposition 35 yields the exact sequence

$$0 \longrightarrow A^{**} \xrightarrow{\;f_1^{**}\;} \Gamma^{*} \xrightarrow{\;f_2^{**}\;} (G/A)^{**} \longrightarrow 0$$

and the commutative diagram

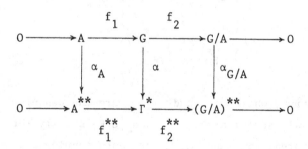

As A is a compactly generated LCA-group and G/A is a discrete group, both A and G/A satisfy the duality theorem and so α_A and $\alpha_{G/A}$ are topological isomorphisms. By Proposition 37, α is an algebraic isomorphism. Since f_1, f_1^{**} and α_A are all open maps and $\alpha f_1 = f_1^{**}\alpha_A$, we see that α is also an open map, and hence a topological isomorphism. //

We can now prove the structure theorem for compactly generated LCA-groups, from which the Principal Structure Theorem for all LCA-groups is a trivial consequence.

Theorem 24. *Let* G *be a compactly generated LCA-group. Then* G *is topologically isomorphic to* $R^a \times Z^b \times K$, *for some compact group* K *and non-negative integers* a *and* b .

Proof. By Proposition 34, we have an exact sequence

$$0 \longrightarrow C \overset{f_1}{\longrightarrow} G \overset{f_2}{\longrightarrow} R^a \times Z^b \times T^c \times F \longrightarrow 0$$

where C is a compact group, F is a finite discrete group and a , b and c are non-negative integers. By Proposition 35 we, therefore, have an exact sequence

$$0 \longleftarrow C^* \overset{f_1^*}{\longleftarrow} G^* \overset{f_2^*}{\longleftarrow} R^a \times T^b \times Z^c \times F \longleftarrow 0$$

where f_2^* is an open map. So G^* has an open subgroup topologically isomorphic to $R^a \times T^b$. As R and T are divisible groups, Proposition 18 says that G^* is topologically isomorphic $R^a \times T^b \times D$, for some discrete group D . As G satisfies the duality theorem G is topologically isomorphic to G^{**} which in turn is topologically isomorphic to $R^a \times Z^b \times K$, where K is the compact group D^* . //

Since every LCA-group has an open compactly generated subgroup we obtain the Principal Structure Theorem.

Theorem 25 (Principal Structure Theorem). *Every LCA-group*
has an open subgroup topologically isomorphic to $R^a \times K$,
for some compact group K *and non-negative integer* a .

As an immediate consequence we have the following signifi-
cant result.

Theorem 26. *Every connected LCA-group is topologically*
isomorphic to $R^a \times K$, *where* K *is a compact connected*
group and $a \geqslant 0$.

Remarks.

(i)   Theorem 24 generalizes the well-known result that
       every finitely generated abelian group is the direct
       product of a finite number of copies of the infinite
       cyclic group with a finite group.

(ii)   One might suspect that one could improve upon the
       Principal Structure Theorem and show that every LCA-
       group is topologically isomorphic to $R^a \times K \times D$,
       where K is compact, D is discrete and $a \geqslant 0$.
       Unfortunately as the following example shows, this
       statement is false.

Example.   Let G be the group $\prod_{i=1}^{\infty} H_i$, where each H_i is
a cyclic group of order four. Let K be the subgroup of
G consisting of all elements $g \in G$ such that $2g = 0$.
Then K is algebraically isomorphic to $\prod_{i=1}^{\infty} K_i$, where
each K_i is a cyclic group of order two. Put the discrete
topology on each K_i and the product topology on K . So
K is a compact totally disconnected topological group.
    Define a topology on G as follows: A base of open
neighbourhoods at 0 in G consists of all the open subsets of K containing 0 . With this topology G is a
totally disconnected LCA-group having K as an open subgroup.

By the Principal Structure Theorem, G has an open subgroup H topologically isomorphic to $R^a \times C$, where C is compact and $a \geqslant 0$. As G is totally disconnected $a = 0$. Suppose that G is topologically isomorphic to $H \times D$, where D is a discrete subgroup of G. As G is not compact D must be infinite. But this is impossible as every infinite subgroup of G has infinitely many elements in K and any discrete subgroup of K must be finite. //

EXERCISE SET THIRTEEN

1. Show that if G is an LCA-group such that G and its dual group are connected, then G is topologically isomorphic to R^n, for some non-negative integer n.

2. Show that an LCA-group G has enough continuous homomorphisms into R to separate points if and only if G is topologically isomorphic to $R^n \times D$, where D is a discrete torsion-free abelian group.
(Hint: Observe that a compact group admits no non-trivial continuous homomorphisms into R.)

3. Describe the compactly generated LCA-groups which are topologically isomorphic to their dual groups.

4. A topological group G is said to be *solenoidal* if there exists a continuous homomorphism f of R into G such that $\overline{f(R)} = G$.

      (i)   Show that if G is also locally compact Hausdorff, then G is either a compact connected abelian group or is topologically isomorphic to R.

      (ii) Show that if G is a compact Hausdorff

solenoidal group then the dual group of G is topologically isomorphic to a subgroup of R_d , the group of reals with the discrete topology.

(Hint for (i): Observe that $\overline{f(R)}$ is topologically isomorphic to $R_1 \times R_2 \times \ldots \times R_n \times K$, where each R_i is a copy of R . Let p_i be the projection of $\overline{f(R)}$ onto R_i and note that $p_i f$ is a continuous homomorphism of R into R_i .)

5. Let F be a field with a topology such that the algebraic operations are continuous. (The additive structure of F , then, is an abelian topological group.) Show that if F is locally compact Hausdorff and connected then F , as a topological group, is isomorphic to R^n , for some $n \geqslant 1$. (A further analysis would show that F is either the real number field R $(n = 1)$, the complex number field $(n = 2)$ or the quaternionic field $(n = 4)$.)

6. Show that if G is any LCA-group then there exists a continuous one-one homomorphism β of G onto a dense subgroup of a compact Hausdorff abelian group. Prove this by two different methods.

(Hint: (1) Use the fact that any LCA-group has enough characters to separate points.

(2) Alternatively, let Γ be the dual group of G and Γ_d the group Γ endowed with the discrete topology. Put $K = (\Gamma_d)^*$ and let β be defined by

$$(g,\gamma) = (\gamma,\beta(g)) , \quad g \in G , \quad \gamma \in \Gamma .$$

The group $K = (\Gamma_d)^*$ is called the *Bohr-compactification* of G .)

7. Let Γ be any LCA-group and $\gamma_1,\ldots,\gamma_n \in \Gamma$. If ϕ is any homomorphism of Γ into T , show that there is a

continuous homomorphism ψ of Γ into T such that $|\psi(\gamma_i) - \phi(\gamma_i)| < \epsilon$, $i = 1,\ldots,n$. (Hint: Use Problem 6, above, method (2).)

* * * * * * * * *

7 · Consequences of the duality theorem

We begin this chapter by showing that the dual of a sub-group is a quotient group and the dual of a quotient group is a subgroup.

Definition. Let H be a closed subgroup of the LCA-group G and Λ the set of all γ in the dual group Γ of G such that $(h,\gamma) = 0$, for all $h \in H$. Then Λ is called the *annihilator* of H .

For fixed $h \in H$, the continuity of (h,γ) shows that the set of all γ with $(h,\gamma) = 0$ is closed, so that Λ is the intersection of closed sets and is therefore closed. Clearly Λ is a group and so it is a closed subgroup of Γ .

Proposition 38. *With the above notation, if Λ is the annihilator of H , then H is the annihilator of Λ .*

Proof. If $h \in H$, then $(h,\gamma) = 0$ for all $\gamma \in \Lambda$. If $g \in G$ and $g \notin H$ then, by Corollary 1 of Theorem 21, there is a $\gamma \in \Lambda$ such that $(g,\gamma) \neq 0$. //

Theorem 27. *Let H be a closed subgroup of an LCA-group G . If Γ is the dual group of G and Λ is the annihilator of H , then Λ and Γ/Λ are topologically isomorphic to the dual groups of G/H and H , respectively.*

Proof. Let f be the canonical homomorphism of G onto G/H . Then Proposition 30 says that the map $f^*: (G/H)^* \to G^*$ is a continuous one-one homomorphism. Further, the last

paragraph of the proof of Proposition 35 shows that f^* is
a homeomorphism of $(G/H)^*$ onto its image in G^* . Of
course the definition of f^* tells us that $f^*((G/H)^*) = \Lambda$,
and so Λ is topologically isomorphic to $(G/H)^*$.

The fact that Γ/Λ is topologically isomorphic to H^*
then follows from the above, together with Proposition 38
and the duality theorem. //

As a corollary we have the following generalization of
Corollary 2 of Theorem 15.

Corollary. *If H is a closed subgroup of an LCA-group G ,
then every character on H can be extended to a character
on G .*

Proof. If ϕ is a character on H , then by the above
theorem $\phi \in \Gamma/\Lambda$. If f is the canonical homomorphism of
Γ onto Γ/Λ and $f(\gamma) = \phi$, $\gamma \in \Gamma$, then $(h,\gamma) = (h,\phi)$,
for all $h \in H$. So γ is the required extension of ϕ . //

We now record an application in the area of
diophantine approximation.

Firstly observe that the definition of the annihilator of
a subgroup H of an LCA-group G would make sense even if
H were not closed in G . However it is obvious that the
annihilator of H , $\Lambda(H)$, would equal $\Lambda(\bar{H})$. We will see
that this observation is quite useful.

Proposition 39. *Let G be a (not necessarily closed)
subgroup of R^n , $n \geqslant 1$. Let $\Lambda(G)$ denote the annihilator
of G in $(R^n)^* = R^n$. Then $\Lambda(G) = \{y \in R^n$ such that
$(y|x)$ is an integer for each $x \in G\}$, where $y = (y_1,\dots,y_n)$
$\in R^n$, $x = (x_1,\dots,x_n) \in G \subseteq R^n$ and $(y|x) = \sum\limits_{i=1}^{n} y_i x_i$.*

Proof. Exercise.

As an immediate consequence of Proposition 38 we have the following result.

Proposition 40. *If G is any subgroup of R^n, $n \geqslant 1$, then $\Lambda(\Lambda(G)) = \bar{G}$.*

Translating Proposition 40 using Proposition 39 we obtain a characterization of those points lying in the closure of a subgroup of R^n.

Proposition 41. *A point x lies in the closure of a subgroup G of R^n, $n \geqslant 1$ if and only if $(y|x)$ is an integer for all $y \in R^n$ such that $(y|g)$ is an integer for all $g \in G$.*

We apply this characterization to the case where G is the subgroup of R^n, $n \geqslant 1$ generated by the vectors $e_1, \ldots, e_n$ of the canonical basis and by an arbitrary number m of points a_i, $i = 1, \ldots, m$ of R^n. To say that $(y|e_i)$ is an integer, for $i = 1, \ldots, n$ means that each coordinate of y is an integer. So we obtain Kronecker's theorem.

Theorem 28 (Kronecker). *Let $a_i = (a_{i1}, \ldots, a_{in})$ for $i = 1, \ldots, m$ and $b = (b_1, \ldots, b_n)$ be points of R^n, $n \geqslant 1$. In order that for each $\varepsilon > 0$ there exists integers $q_1, \ldots, q_m$ and integers $p_1, \ldots, p_n$ such that*

$$\left| q_1 a_{1j} + q_2 a_{2j} + \ldots + q_m a_{mj} - p_j - b_j \right| \leqslant \varepsilon \quad \text{for} \quad j = 1, \ldots, n$$

it is necessary and sufficient that for each finite sequence $r_1, \ldots, r_n$ of integers such that the numbers $\sum_{j=1}^{n} a_{ij} r_j$ for

$i = 1,\ldots,m$ *are all integers, the number* $\sum_{j=1}^{n} b_j r_j$ *should also be an integer.*

Putting $m = 1$ we obtain the following corollary which is a generalization of Problem 3(ii) of Exercise Set Five.

Corollary. *Let* $\theta_1,\ldots,\theta_n$ *be real numbers. In order that, given any* n *real numbers* $x_1,\ldots,x_n$ *and a real number* $\varepsilon > 0$ *, there should exist an integer* q *and* n *integers* p_j *such that*

$$|q\theta_j - p_j - x_j| \le \varepsilon \quad \text{for} \quad j = 1,\ldots,n$$

it is necessary and sufficient that there exist no relation of the form $\sum_{j=1}^{n} r_j \theta_j = h$ *, where the* r_j *are integers not all zero and* h *is an integer. (In particular this implies that the* θ_j *and the ratios* θ_j/θ_k *,* $j \ne k$ *, must be irrational.)*

For some further comments in this area of approximation see Section 26.19 of E. Hewitt and K.A. Ross, Abstract Harmonic Analysis I and Chapter VII, Section 1.3 of N. Bourbaki, Elements of Mathematics, General Topology II.

EXERCISE SET FOURTEEN

1. Prove that a closed subgroup of a compactly generated LCA-group is compacted generated.
(Hint: Use Corollary 2 of Theorem 7.)

2. (i) Show that any closed subgroup of $R^n \times T^m$ is topologically isomorphic to $R^a \times Z^b \times T^c \times D$, where D is a discrete finite group, $a + b \le n$ and $c \le m$.

(ii)   Show that any Hausdorff quotient group of
$R^n \times T^m$ is topologically isomorphic to $R^a \times T^b$
with $a \leqslant n$ and $a + b \leqslant n + m$.

(iii) Show that any closed subgroup of $R^n \times T^m \times F$,
where F is a discrete free abelian group, is
topologically isomorphic to $R^a \times Z^b \times T^c \times D \times F'$,
where $a + b \leqslant n$, $c \leqslant m$, D is a discrete
finite group and F' is a subgroup of F .

(iv)   Show that any closed subgroup of $R^n \times T^m \times D$,
where D is a discrete abelian group, is topologically isomorphic to $R^a \times T^b \times D'$, where D'
is a discrete group, $a \leqslant n$ and $b \leqslant m$.

(v)   Show that any Hausdorff quotient of $R^n \times T^m \times D$,
where D is a discrete abelian group, is topo-
logically isomorphic to $R^a \times T^b \times D'$, where D'
is a discrete group, $a \leqslant n$ and $a + b \leqslant n + m$.

(vi)   Hence show that any closed subgroup or Hausdorff
quotient of $R^n \times Z^m \times K$, where K is a compact
Hausdorff abelian group, is topologically isomorphic to $R^a \times Z^b \times K'$, where K' is a compact
group, $a \leqslant n$ and $a + b \leqslant n + m$.

3. Let G be a compactly generated LCA-group. Assuming
the structural result (Proposition 33) which says that G
has a subgroup A topologically isomorphic to Z^n , $n \geqslant 0$,
such that G/A is compact, show that G^* is locally isomorphic to R^n . Hence prove that G is topologically
isomorphic to $R^a \times Z^b \times K$, where $a + b = n$ and K is
compact.

4. Let G be any LCA-group. Assuming the structural
result (Theorem 26) which says that every connected LCA-
group is topologically isomorphic to $R^a \times K$, for some
compact group K and $a \geqslant 0$, prove the Principal Structure

Theorem. (Hint: Let C be the component of 0 in G and observe that Problem 4(i) of Exercise Set 3 says that G/C has a compact open subgroup A . So G has an open subgroup H with the property that H/C is compact. Deduce that H^* is locally isomorphic to R^n .)

5. Let G be a compactly generated LCA-group. Assuming the Principal Structure Theorem show that G is topologically isomorphic to $R^a \times Z^b \times K$, where K is compact, $a \geq 0$ and $b \geq 0$. (Hint: Observe that G^* has an open subgroup topologically isomorphic to $R^a \times K_1$, where K_1 is compact and $a \geq 0$. Noting that K_1^* is a quotient group of G we see that it must be finitely generated and so K_1 is topologically isomorphic to $T^b \times F$, where F is a finite discrete group and $b \geq 0$.)

6. Let G be a subgroup of R^n , $n \geq 1$ and $\Lambda(G)$ its annihilator. Show that $\Lambda(G) = \{y \in R^n$ such that $(y|x)$ is an integer for each $x \in G\}$, where $y = (y_1,\ldots,y_n) \in R^n$, $x = (x_1,\ldots,x_n) \in G \subseteq R^n$ and $(y|x) = \sum_{i=1}^n y_i x_i$. (Hint: Recall that $T = R/Z$ and use Example 3 in Chapter 3 and Theorem 13.)

7. If G is an LCA-group, C is the component of 0 in G and G/C is compact show that G is topologically isomorphic to $R^n \times K$, for some compact group K and $n \geq 0$.

* * * * * * * * *

Theorem 29. *Let* G *be an LCA-group and* Γ *its dual group. Then* G *is metrizable if and only if* Γ *is σ-compact.*

Proof. Assume that G is metrizable. Then G has a countable base of compact neighbourhoods $U_1, U_2, \ldots$ of 0 .

By Theorem 11, if $a = \frac{1}{5}$, then the sets $\overline{P(U_i,V_a)}$,
$i = 1,2,\ldots,$ are compact neighbourhoods of 0 in Γ .
As each $\gamma \in \Gamma$ is continuous, $\Gamma = \bigcup_{i=1}^{\infty} \overline{P(U_i,V_a)}$. So Γ
is σ-compact.

Conversely, assume that Γ is σ-compact. By Problem 4
of Exercise Set Twelve, there exists a family $\{Y_n\}$,
$n = 1,2,\ldots$ of compact neighbourhoods of 0 in Γ such
that every compact subset of Γ lies in some Y_n and
$Y_n \subseteq Y_{n+1}$, $n \geqslant 1$. So the family $\{g \in G: (g,\gamma) \in V_{1/k}$,
for all $\gamma \in Y_n\}$, for $k = 2,3,\ldots$ and $n = 1,2,\ldots$ is a
base of neighbourhoods of 0 in G . (Observe that in
saying that these sets are neighbourhoods we are using the
fact that G is the dual group of Γ .) Thus G has a
countable base of neighbourhoods at 0 which, by Problem
3(ii) of Exercise Set Six, implies that G is metrizable. //

As a corollary we have the following striking result.

Corollary. *Let G be an LCA-group. Then G is compact
and metrizable if and only if Γ is countable.*

Proof. Exercise.

The remainder of this chapter is devoted to characterizing
those compact groups which are connected.

Proposition 42. *Let G be a locally compact Hausdorff
group and K a compact subset of G . Then there exists
an open and closed compactly generated subgroup of G containing K .*

Proof. Exercise.

Definition. An element g of a topological group G is said to be *compact* if $\overline{\text{gp}\{g\}}$, the smallest closed subgroup of G containing g , is compact.

Example. An element $g \in G = R^a \times Z^b \times K$, where K is compact and $a \geqslant 0$ and $b \geqslant 0$, is compact if and only if it lies in $\{0\} \times \{0\} \times K$.

Proposition 43. *If G is an LCA-group then the set S of compact elements is a closed subgroup of G .*

Proof. If g and h are in S , then $g - h \in \text{gp}\{g\} + \text{gp}\{h\} \subseteq \overline{\text{gp}\{g\}} + \overline{\text{gp}\{h\}}$. As $g - h$ lies in the compact group $\overline{\text{gp}\{g\}} + \overline{\text{gp}\{h\}}$, it is compact and so is in S . Thus S is a subgroup of G .

Now let $x \in \overline{S}$. By Proposition 42 there is an open compactly generated subgroup H of G containing x . By Theorem 24, H is topologically isomorphic to $R^a \times Z^b \times K$, where K is compact, $a \geqslant 0$ and $b \geqslant 0$. To see that $x \in S$ we only have to show that the R^a coordinate of x and the Z^b coordinate of x are both zero. If the R^a coordinate of x or the Z^b coordinate of x were different from zero, then there would be an entire neighbourhood of x disjoint from S — since all the compact elements of $R^a \times Z^b \times K$ lie in $\{0\} \times \{0\} \times K$. This would contradict the fact that $x \in \overline{S}$. //

Remark. The set of compact elements of a non-abelian locally compact Hausdorff group need not be a subgroup of G . As an example of this, let G be the discrete group generated by two elements a and b with the relations $a^2 = b^2 = e$. Then a and b are compact elements, but ab is not compact.

Theorem 30. *Let G be an LCA-group, Γ its dual group,*

S *the set of compact elements in* Γ *and* C *the component of* 0 *in* G . *Then* S *is the annihilator in* Γ *of* C *and* C *is the annihilator in* G *of* S .

Proof.

(i)   Suppose that there is a compact subgroup K of Γ such that $K \neq \{0\}$. Then G has a discrete quotient group K^* . As K^* is not connected, G is not connected.

(ii) Suppose that G is totally disconnected. Let $\gamma \in \Gamma$ and U be a neighbourhood of 0 in G such that $\gamma(g) \in V_{\frac{1}{4}}$, for all $g \in U$. As G is locally compact and totally disconnected, Problem 4(i) of Exercise Set Three implies that U contains a compact open subgroup K . It is clear that $\gamma(K)$ is a subgroup of $V_{\frac{1}{4}}$ and hence equals $\{0\}$; that is, $\gamma \in \Lambda(\Gamma,K)$ the annihilator in Γ of K . As K is open in G , $\Lambda(\Gamma,K)$, which is the dual group of the discrete group G/K , is compact. So every $\gamma \in \Gamma$ lies in a compact group. Hence $S = \Gamma$.

(iii) Now let G be any LCA-group. Then G/C is a totally disconnected LCA-group and so, by (ii), the dual group of G/C contains only compact elements. As $(G/C)^*$ is topologically isomorphic to $\Lambda(\Gamma,C)$, we see that $\Lambda(\Gamma,C) \subseteq S$. As C is a connected LCA-group (i) implies that C^* has no non-trivial compact subgroup; that is, $\Gamma/\Lambda(\Gamma,C)$ has no compact subgroup. If $\gamma \in S$ and $\gamma \notin \Lambda(\Gamma,C)$ then there is a compact subgroup K of Γ whose image under the canonical map $f \colon \Gamma \to \Gamma/\Lambda(\Gamma,C)$ is not $\{0\}$. This image is a compact subgroup of $\Gamma/\Lambda(\Gamma,C)$ and so we have a contradiction. Hence $S \subseteq \Lambda(\Gamma,C)$. Thus $S = \Lambda(\Gamma,C)$.

The dual statement then follows from Proposition 38. //

Corollary 1.   *Let G be an LCA-group. Then the following*
properties are equivalent:
(i)   G is totally disconnected
(ii) every element in G^* is compact.

Corollary 2.   *Let G be an LCA-group. Then G is*
connected if and only if G^* has no compact subgroup $\neq \{0\}$.

Corollary 3.   *Let G be a compact Hausdorff abelian group*
with C the component of 0 in G . Let Φ be the torsion
subgroup of Γ (that is, the subgroup consisting of all
elements of finite order). Then $\Phi = \Lambda(\Gamma,C)$, the annihilator
in Γ of C , and $C = \Lambda(G,\Phi)$. Also Φ is isomorphic to
$(G/C)^*$.

Proof.   The first two equalities follow from the theorem
above, since an element of a discrete group is compact if
and only if it has finite order. The last statement follows
from the duality between subgroups and quotients. //

As an immediate consequence of Corollary 3 we have an
interesting characterization of those compact groups which
are connected.

Corollary 4.   *A compact Hausdorff abelian group is connected*
if and only if its dual group is torsion free.

Notation.   Let G be an abelian group and f_n the homomorphism of G into itself given by $f_n(g) = ng$, for n
a positive integer. We denote $f_n(G)$ by $G^{(n)}$ and $f_n^{-1}\{0\}$
by $G_{(n)}$.

Proposition 44.   *Let G be an LCA-group with dual group*
Γ . For any n > 0 ,

(i) $\Lambda(\Gamma, G^{(n)}) = \Gamma_{(n)}$

and (ii) $\Lambda(\Gamma, G_{(n)}) = \overline{\Gamma^{(n)}}$

Proof. Let $\gamma \in \Lambda(\Gamma, G^{(n)})$. Then for each $g \in G$ we have $ng \in G^{(n)}$ and so $\gamma(ng) = n\gamma(g) = 0$. Thus $\gamma \in \Gamma_{(n)}$. Conversely, if $\gamma \in \Gamma_{(n)}$, then $n\gamma(g) = \gamma(ng) = 0$, for all $g \in G$ and so $\gamma \in \Lambda(\Gamma, G^{(n)})$. Hence (i) is true.

To prove (ii) regard G as the dual group of Γ . Then, by (i), $\Lambda(G, \Gamma^{(n)}) = G_{(n)}$ and so

$$\overline{\Gamma^{(n)}} = \Lambda(\Gamma, \Lambda(G, \overline{\Gamma^{(n)}})) = \Lambda(\Gamma, \Lambda(G, \Gamma^{(n)})) = \Lambda(\Gamma, G_{(n)}) . \; /\!/$$

Theorem 31. *Let G be an LCA-group and Γ its dual group. If G is divisible then Γ is torsion-free. If Γ is torsion-free then $G^{(n)}$ is dense in G , for n = 1,2,... If G is discrete or compact, then G is divisible if and only if Γ is torsion-free.*

Proof. If G is divisible then $G^{(n)} = G$, for all n and so Proposition 44(i) implies that $\Gamma_{(n)} = \{0\}$; that is, Γ is torsion-free.

If Γ is torsion-free then Proposition 44(ii) shows that $\Lambda(G, \Gamma_{(n)}) = G = \overline{G^{(n)}}$, for each $n \geqslant 1$.

If G is discrete then $G^{(n)}$ is certainly closed, for each $n \geqslant 1$. If G is compact then $G^{(n)}$ is a continuous image of G and so is compact and closed. So in both cases $G = G^{(n)}$, for all $n \geqslant 1$, and G is divisible . $/\!/$

Corollary 1. *Let G be a compact Hausdorff abelian group and Γ its dual group. Then the following are equivalent:*
(i) G is connected.
(ii) Γ is torsion-free.
(iii) G is divisible.

Corollary 2. *Every connected LCA-group* G *is divisible.*

Proof.   G is topologically isomorphic to $R^n \times K$, where K
is compact and connected, from which the result immediately
follows. //

Remarks.   It is also true that a compact Hausdorff non-
abelian group is connected if and only if it is divisible.
(See J. Mycielski, Some properties of connected compact
groups, *Colloq. Math.* 5 (1958) 162-166.)
    Corollary 2 above does not extend to the non-abelian
case. (See Section 24.44 of E. Hewitt and K.A. Ross,
Abstract Harmonic Analysis I.)

EXERCISE SET FIFTEEN

1. If G is an LCA-group, show that it is compact and
metrizable if and only if its dual group is countable.
(Hint: Use the Corollary of Theorem 2.)

2. Show that the Bohr-compactification bG of an LCA-
group is metrizable if and only if G is compact and
metrizable and bG = G .
(Hint: See Problem 6 of Exercise Set Thirteen and use
Problem 1, above.)

3. Let K be a compact subset of a locally compact
Hausdorff group G . Show that there exists an open and
closed compactly generated subgroup of G containing K .

4. Let G be an LCA-group and Γ its dual group. Let
a be the least cardinal number of an open basis at O in
G and b be the least cardinal number of a family of
compact subsets of Γ whose union is Γ . Show that a = b .

(Hint: To prove $a \leqslant b$, let $S = \{A_i : i \in I\}$ be a family of compact sets such that $\Gamma = \underset{i \in I}{\cup} A_i$ and b equals the cardinality of the index set I . Let B_i be an open set containing A_i such that $\bar{B}_i$ is compact, for each $i \in I$. Let S' be the family of all finite unions of sets $\bar{B}_{i_1} \cup \bar{B}_{i_2} \cup \ldots \cup \bar{B}_{i_n}$. Verify that Γ is the union of the members of S' and that every compact subset of Γ is a subset of a member of S' . Now proceed as in the proof of Theorem 29.)

5. (i)   Let G be a compact Hausdorff abelian group and $w(G)$ the least cardinal number of an open basis of G . Show that $w(G)$ equals the cardinal number of G^* .
        (Hint: Use Problem 4 above.)

   (ii) If G is a compactly generated LCA-group, show that $w(G) = w(G^*)$.

   (iii) If G is any LCA-group, show that $w(G) = w(G^*)$.
        (Hint: If G is finite, the result is trivial, so assume G is infinite. Then G has an open subgroup H topologically isomorphic to $R^n \times K$, where K is compact and $n \geqslant 0$. Show that
        $$w(G) = \max\left[w(R^n \times K), \text{cardinal number of } G/H\right] .$$
        Observing that G^* has a compact subgroup A , topologically isomorphic to $(G/H)^*$ such that G^*/A is topologically isomorphic to $R^n \times K^*$, show that
        $$w(G^*) \geqslant \max\left[w(R^n \times K^*), w((G/H)^*)\right] .)$$

6. Let G be a compact Hausdorff abelian group with dual group Γ . Show that the following conditions are equivalent (where c denotes the cardinal number of R)
(i) G is solenoidal

102

(ii) Γ is algebraically isomorphic to a subgroup of R

(iii) Γ is torsion-free and the cardinal number of $\Gamma \leqslant c$

(iv)   G is connected and $w(G) \leqslant c$.

(Hint: See Problem 4 of Exercise Set Thirteen. Assume the
fact, from abelian group theory, that (iii) implies (ii).)

7. Let G be a divisible compact Hausdorff (not
necessarily abelian) group. Prove that G is connected.
(Hint: Assume that G is not connected and arrive at a
contradiction by showing that G has a proper open normal
subgroup H , such that G/H is a finite divisible group.)

* * * * * * * * *

8 · Locally Euclidean and NSS-groups

Definition. A topological group G is said to have *no small subgroups*, or to be an *NSS-group*, if there exists a neighbourhood U of e which contains no subgroup other than $\{e\}$.

As an immediate consequence of Corollary 2 of Theorem 14 we have a complete description of compact Hausdorff abelian NSS-groups.

Proposition 45. *Every compact Hausdorff abelian NSS-group is topologically isomorphic to* $T^n \times D$, *for some discrete group* D *and* $n \geq 0$.

The above proposition allows us to describe all locally compact Hausdorff abelian NSS-groups.

Theorem 32. *Every LCA-group* G *which has no small subgroups is topologically isomorphic to* $R^a \times T^b \times D$, *where* D *is some discrete group, and* a *and* b *are non-negative integers.*

Proof. By the Principal Structure Theorem G has an open subgroup topologically isomorphic to $R^a \times K$, for some compact group K and $a \geq 0$. As every subgroup of an NSS-group is an NSS-group, K is an NSS-group. Proposition 45 then implies that K is topologically isomorphic to $T^b \times S$, where S is a discrete group and $b \geq 0$. So G has an open subgroup topologically isomorphic to $R^a \times T^b$. Proposition 18 then shows that G is topologically isomorphic to $R^a \times T^b \times D$, for some discrete group D . //

Remark.   In 1900 David Hibert presented to the International
Congress of Mathematicians in Paris a series of 23 research
projects (*Bull. Amer. Math. Soc.* 8 (1901) 437–479).   The
spirit of his fifth problem is: What *topological* conditions
on a topological group will ensure that the group admits an
analytical structure which makes it into a Lie group?
(Roughly speaking a topological group is said to be a *Lie
group* if the component of the identity is open and it has
the additional structure of a differentiable manifold with
the operations $(x,y) \to xy$ and $x \to x^{-1}$ being analytic.
As examples we mention R^n , T^n and discrete groups.)   In
particular he asked if a locally Euclidean group is a Lie
group.   An affirmative answer was given in 1952 by A. Gleason,
D. Montgomery and L. Zippin.   Another formulation is: a
locally compact group is a Lie group if it has no small subgroups.   Theorem 32 above is the abelian case of this theorem.

The remainder of this chapter is devoted to proving that
a locally Euclidean *abelian* topological group is topologically
isomorphic to $R^a \times T^b \times D$, where D is discrete, $a \geqslant 0$ and
$b \geqslant 0$, and so is a Lie group.

For a full discussion of Hibert's fifth problem see D.
Montgomery and L. Zippin, Topological transformation groups,
and I. Kaplansky, Lie algebras and locally compact groups.

The reader may also be interested in J. Szenthe, "Topological characterization of Lie group actions", *Acta
Scientiarum Mathematicarum* 36 (1974) 323–344, where it is
shown that a locally compact group is a Lie group if and only
if it is locally contractible.

Definition.   A topological space X is said to be *locally
connected* if for each $x \in X$ and each open neighbourhood
U of x , there is a connected neighbourhood V of x
such that $V \subseteq U$.

Definition. A topological space X is said to be *locally Euclidean* if each $x \in X$ has a neighbourhood U homeomorphic to the unit sphere in R^n, for some $n \geqslant 0$.

Examples. All Lie groups are locally Euclidean; for example, any group of the form $R^a \times T^b \times D$, where D is discrete, $a \geqslant 0$ and $b \geqslant 0$.

Of course every locally Euclidean space is locally connected.

Theorem 33. *Every locally connected LCA-group* G *is topologically isomorphic to* $R^a \times K \times D$, *where* K *is some compact connected locally connected group,* D *is some discrete group and* $a \geqslant 0$.

Proof. Let C be the component of 0 in G. As G is locally connected, C is an open subgroup of G. By Corollary 2 of Theorem 31, C is divisible. So G is topologically isomorphic to $C \times D$, where D is the discrete group G/C. As C is connected it is topologically isomorphic to $R^a \times K$, for some compact connected group K and $a \geqslant 0$. So G is topologically isomorphic to $R^a \times K \times D$. As G is locally connected and K is a quotient group of G, K is locally connected. //

Remark. The above theorem reduces the problem of characterizing the locally connected LCA-groups to that of characterizing the compact connected locally connected Hausdorff abelian groups.

Remarks on dimension. Let X be a set and Σ a finite family of subsets of X. For each $x \in X$ we denote by $m(x)$ the number of sets $S \in \Sigma$ such that $x \in S$. The *multiplicity* $m(\Sigma)$ is defined as $\max\{m(x) : x \in X\}$. A family Σ' of subsets of X is said to *refine* Σ if for every $S' \in \Sigma'$ there is an $S \in \Sigma$ such that $S' \subseteq S$.

Now let X be a compact Hausdorff space and n a non-negative integer. Then X is said to have *dimension* n if the following two conditions are satisfied: (i) every finite covering of X by open sets admits a finite covering by closed sets which refines the given open cover and has multiplicity $\leqslant n+1$; (ii) there exists a finite covering of X by open sets such that every finite covering by closed sets which refines the given open cover has multiplicity $> n$. (If no such n exists, X is said to have infinite dimension.) We write $\dim(X) = n$.

Theorem. *The dimension of the unit sphere in* R^n *is* n .

The proof of this theorem is non-trivial - see, for example, J. Nagata, *Modern Dimension Theory*, North Holland, 1965.

We need one more fact about dimension, the proof of which is quite easy.

Proposition. *Let* Y *be a closed subspace of a compact Hausdorff space* X . *Then* $\dim(Y) \leqslant \dim(X)$.

Remarks on rank. A finite subset $\{x_1,\ldots,x_n\}$ of a torsion-free abelian group G is said to be *linearly independent* if $m_1 x_1 + \ldots + m_n x_n = 0$, for integers $m_1,\ldots,m_n$, implies that $m_1 = m_2 = \ldots = m_n = 0$. The linearly independent set $\{x_1,\ldots,x_n\}$ is said to be *maximal* in G if, for each $x \in G\backslash\{x_1,\ldots,x_n\}$, the set $\{x,x_1,\ldots,x_n\}$ is not linearly independent. It can be shown that if G has a maximal linearly independent set $\{x_1,\ldots,x_n\}$ then no linearly independent subset of G has more than n elements. So we can define the *rank* of G to be the number of elements in a maximal linearly independent subset. If G has linearly

independent subsets with arbitrarily large numbers of
elements, then G is said to have *infinite rank*. We will
have cause to use the following easily verified proposition.

Proposition. *Let G be a torsion-free abelian group of
infinite rank. Then for each natural number n , G has a
quotient group H which is a torsion-free abelian group of
rank n .*

Proposition 46. *Let G be a discrete torsion-free abelian
group of rank n . Then G^* has a subspace homeomorphic to
$\prod_{i=1}^{n} X_i$, where each X_i is a homeomorphic copy of the open
unit interval (0,1) .*

Proof. Let $\{g_1,\ldots,g_n\}$ be a maximal linearly independent
subset of G . For each $t = (t_1,\ldots,t_n) \in \prod_{i=1}^{n} X_i$ we define
a character γ_t on G as follows: If $g \in G$, then there
exist integers $m,m_1,\ldots,m_n$ such that $mg = m_1 g_1 + \ldots + m_n g_n$,
so we put

$$\gamma_t(g) \;=\; \exp\left[2\pi i\left(\tfrac{m_1}{m} t_1 + \ldots + \tfrac{m_n}{m} t_n\right)\right] .$$

That γ_t is indeed a character is easily verified. The map
of $\prod_{i=1}^{n} X_i$ into G^* given by $t \to \gamma_t$ is clearly one-one
and a routine verification shows that it maps $\prod_{i=1}^{n} X_i$
homeomorphically onto its image in G^* . //

Theorem 34. *Let G be a compact connected Hausdorff
abelian group. Then G has infinite dimension if and only
if G^* has infinite rank. If G has finite dimension,
then the dimension of G is equal to the rank of G^* .*

Proof. Note that it is implicit in the statement of the
theorem that G^* is torsion-free, of course we proved this

108

in Corollary 4 of Theorem 30.

First we show that $\dim(G) \geqslant \mathrm{rank}(G^*)$. If G^* has rank n then Proposition 46 says that G has a subspace homeomorphic to $\prod_{i=1}^{n} X_i$, where $X_i = (0,1)$. So $\dim(G) \geqslant n = \mathrm{rank}(G^*)$. If G^* has infinite rank, then it has a quotient group H which is torsion-free of rank n , for each natural number n . So G has a subgroup topologically isomorphic to H^* which in turn has a subspace homeomorphic to $\prod_{i=1}^{n} X_i$ Hence $\dim(G) \geqslant n$, for each n ; that is, G has infinite dimension.

Now we show that $\mathrm{rank}(G^*) \geqslant \dim(G)$. Assume that $\dim(G) \geqslant n$, for some natural number n . Let Σ be a finite open cover of G such that any closed cover that refines Σ has multiplicity $> n$. For each $g \in G$ there is a neighbourhood V_g of 0 such that $g + 2V_g$ is contained in some $S \in \Sigma$. As G is compact, a finite number of the sets $g + V_g$ cover G ; that is, $G = (g_1 + V_{g_1}) \cup ... \cup (g_m + V_{g_m})$. Put $V = V_{g_1} \cap ... \cap V_{g_m}$. By Corollary 2 of Theorem 14, V contains a closed subgroup H such that G/H is topologically isomorphic to T^k , for some $k \geqslant 0$. (Observe that G/H is connected and so the discrete group in this Corollary must be trivial.) If f is the canonical map of G onto G/H then it is easily verified that $f^{-1}\{y\}$ is a subset of some $S \in \Sigma$, for each $y \in G/H$. So Problem 1 of Exercise Set Sixteen implies that $\dim(G/H) \geqslant n$; that is, $\dim(T^k) = k \geqslant n$. Of course $(G/H)^*$ is topologically isomorphic to Z^k and to a subgroup of G^* . Hence $\mathrm{rank}(G^*) \geqslant k \geqslant n$. So $\mathrm{rank}(G^*) \geqslant \dim(G)$. //

Remark. If G is any abelian group then we define the torsion-free rank of G to be the number of elements in a maximal linearly independent subset of G . The argument in the above theorem then shows that for any compact Hausdorff abelian group G , the dimension of G is equal to the torsion-free rank of G^* .

Theorem 35. *Let* G *be a discrete torsion-free abelian group of finite rank. Then* G^* *is locally connected if and only if* G *is finitely generated.*

Proof. If G is finitely generated then it is topologically isomorphic to $Z^a \times F$, where F is a finite discrete group and a is a non-negative integer. So G^* is topologically isomorphic to $T^a \times F$, and hence is locally connected.

Let G^* be locally connected and suppose that G is not finitely generated. Let $S = \{g_1, \ldots, g_n\}$ be a maximal linearly independent subset of G and W the sub-basic neighbourhood $P(S, V_{\frac{1}{4}})$. Let H be the subgroup generated by S and $A = \Lambda(H)$, the annihilator in G^* of H . We shall show that W is homeomorphic to $R^n \times A$.

For each $t = (t_1, \ldots, t_n) \in \prod_{i=1}^{n} X_i$, where X_i is the open interval $(-\frac{1}{4}, \frac{1}{4})$, we define a character γ_t on G as follows: If $g \in G$ then there exist integers $m, m_1, \ldots, m_n$ such that $mg = m_1 g_1 + \ldots + m_n g_n$, so we put

$$\gamma_t(g) = \exp\left[2\pi i(\underset{\overline{m}}{m_1}t_1 + \ldots + \underset{\overline{m}}{m_n}t_n)\right] .$$

As in Proposition 46, γ_t is a character and the map of $\prod_{i=1}^{n} X_i$ into G^* , given by $t \to \gamma_t$, is a homeomorphism of $\prod_{i=1}^{n} X_i$ onto its image E in G^* . Of course E is homeomorphic to R^n . Now let $\gamma \in W$. Put $t_i = \gamma(g_i)$, $i = 1, \ldots, n$, and $t = (t_1, \ldots, t_n)$. Clearly $\gamma - \gamma_t \in A = \Lambda(H)$. Thus $\gamma = \gamma_t + \eta$, where $\eta \in A$. It is routine to verify that the map of W into $E \times A$ given by $\gamma \to (\gamma_t, \eta)$ is a homeomorphism. In particular, we obtain a continuous open map of W onto A . So we will have our contradiction, and hence the required result, if we can show that A is not locally connected.

110

Since H is finitely generated while G is not, it
follows from Problem 5(ii) of Exercise Set Twelve that
G/H is not finitely generated. So G/H is an infinite
torsion group. By Corollary 3 of Theorem 30, the dual
group of (G/H) is totally disconnected. However A is
topologically isomorphic to $(G/H)^*$ and hence is an
infinite totally disconnected group. Finally observe that
a totally disconnected group is not locally connected
unless it is discrete, but A is not discrete as it is
infinite and compact. So we have the required contradiction. //

We now have as an immediate consequence of Theorems 33,
34 and 35 the main result of this chapter.

Theorem 36. *Let G be a finite-dimensional locally
connected LCA-group. Then G is topologically isomorphic
to $R^a \times T^b \times D$, where D is a discrete group and a and
b are non-negative integers.*

Corollary. *Let G be a locally Euclidean abelian topological group. Then G is topologically isomorphic to
$R^a \times T^b \times D$, where D is a discrete group and a and b
are non-negative integers.*

Remark. J. Dixmier (Quelques proprietés des groupes
abéliens localement compacts, *Bull. Sci. Math.* 81 (1957)
38-48) has characterized the compact connected locally
connected Hausdorff abelian groups as those having dual
groups which are discrete torsion-free and have every subgroup of finite rank free. From this L.S. Pontryagin,
Topological Groups, derives the following generalization
of Theorem 36.

Theorem. *Every locally connected metrizable LCA-group is topologically isomorphic to* $R^a \times D \times \prod_{i=1}^{\infty} T_i$ *or* $R^a \times D \times T^b$, *where each* T_i *is a copy of* T , D *is a discrete group, and* a *and* b *are non-negative integers.*

Unfortunately the structure of non-metrizable locally connected groups is not so pleasant. Dixmier showed that a compact connected locally connected LCA-group need not be path connected and hence need not be topologically isomorphic to a product of copies of T . Some further information is given in Ky Fan's paper, On local connectedness of LCA-groups, *Math. Ann.* 187 (1970) 114-116.

EXERCISE SET SIXTEEN

1. Let X be a compact Hausdorff space of dimension $\geqslant n$ and Σ a finite open cover of X such that every closed cover of X which refines Σ has multiplicity $> n$. If f is a continuous map of X onto a Hausdorff space Y such that, for each $y \in Y$, $f^{-1}\{y\}$ is a subset of a member of Σ , show that $\dim(Y) \geqslant n$.

2. (i) Show that an LCA-group is compactly generated if and only if its dual group has no small subgroups.

    (ii) Deduce that any closed subgroup or Hausdorff quotient group of an LCA-NSS-group is an LCA-NSS-group.

3. Show that a finite product of NSS-groups is an NSS-group, but that an infinite product of NSS-groups is an NSS-group only if it is, in some sense, trivial.

4. Using Problem 2(i) above, show that every LCA-group

G has a compact subgroup K such that G/K is an LCA-NSS-group.

5. (i) Let X be a topological space and d a metric on the *set* X. Then d is said to be a *continuous metric* on X, if the map d: $X \times X \to R$ is continuous, where $X \times X$ denotes the product of two copies of the topological space X. We say, then, that the topological space X *admits a continuous metric*. Show that any compact space which admits a continuous metric is metrizable.

   (ii) Show that any locally compact group which admits a continuous metric is metrizable.
   (Hint: Use (i) and Problem 3(ii) of Exercise Set Six.)

   (iii) If G is a topological group which has a family $V_1, V_2, \ldots$ of neighbourhoods of e such that $\bigcap_{i=1}^{\infty} V_n = \{e\}$, show that G admits a continuous metric.
   (Hint: See Problem 3(i) of Exercise Set Six.)

   (iv) Deduce that a locally compact group is metrizable if and only if it has a countable family $V_1, V_2, \ldots$ of neighbourhoods of e such that $\bigcap_{i=1}^{\infty} V_n = \{e\}$.

   (v) Show that every locally compact NSS-group is metrizable.

6. (i) Let G be a non-discrete locally compact NSS-group. Show that there exists a compact neighbourhood V of e such that for all x and y in V, the relation $x^2 = y^2$ implies $x = y$.
   (Hint: Assume G is not abelian. If the result were false there would exist two sequences

113

$x_1, x_2, \ldots$ and $y_1, y_2, \ldots$ of elements of G both converging to e such that $x_n^2 = y_n^2$ and $x_n^{-1} y_n^{-1} = z_n \neq e$. Let U be a symmetric compact neighbourhood of e not containing any subgroup of G other than $\{e\}$ and let p_n be the smallest integer $p > 0$ such that $(z_n)^{p+1} \notin U$. Show, by passing to a subsequence that we can assume that $z = \lim_{n \to \infty} (z_n)^{p_n}$ exists, is not equal to e, and belongs to U. Show that $z = z^{-1}$ and so obtain a contradiction.)

(ii) Let U be a compact symmetric neighbourhood of e containing no subgroup other than $\{e\}$ and let V be a neighbourhood of e. Prove that there exists a number $c(V) > 0$ such that whenever p and q are positive integers such that $p \leqslant c(V)q$ and $x \in G$ is such that $x, x^2, \ldots, x^q$ are in U then $x^p \in V$.

(Hint: Suppose that there exist sequences $p_1, p_2, \ldots$ and $q_1, q_2, \ldots$ of positive integers such that $\lim_{n \to \infty} (p_n/q_n) = 0$ and for each n, an element $g_n \in G$ such that $(g_n)^k \in U$ for $1 \leqslant k \leqslant q_n$ but $(g_n)^{p_n} \notin V$. Suppose also that the sequence $(g_n)^{p_n}$ has a limit $g \neq e$ such that $g \in U$. Show that $g^m \in U$, for all $m > 0$ and so obtain a contradiction.)

(iii) Let G and V be as in (i). If $a_1, a_2, \ldots$ is any sequence of points of V with e as limit, show that there exists a subsequence $b_1, b_2, \ldots$ of $a_1, a_2, \ldots$ and a sequence $k_1, k_2, \ldots$ of positive integers such that the sequence $b_1^{k_1}, b_2^{k_2}, \ldots$ converges to a point other than e.

(Hint: Consider the smallest of the positive integers k such that $(a_n)^{k+1} \notin V$.)

(iv) Show that if r and s are real numbers such

114
114

that the sequences $b_1^{[rk_1]}, b_2^{[rk_2]}, \ldots$ and $b_1^{[sk_1]}, b_2^{[sk_2]}, \ldots$ converge to x and y respectively, in G , then the sequence $b_1^{[(r+s)k_1]}, b_2^{[(r+s)k_2]}, \ldots$ converges to xy . (Here $[t]$ denotes the integer part of the real number t .)

(v) Using (iii) and (iv) show that for every dyadic number $r \in [0,1]$ the sequence $b_1^{[rk_1]}, b_2^{[rk_2]}, \ldots$ converges in G .

(vi) If W is any neighbourhood of e such that $W \subseteq V$, show that for every real number $r \in [0,1]$ there is a dyadic number s such that $b_n^{[(r+s)k_n]} \in W$, for all n . Deduce that for each $r \in [0,1]$, the sequence $b_1^{[rk_1]}, b_2^{[rk_2]}, \ldots$ has a limit f(r) . (Hint: Use (ii).)

(vii) If $-1 \leqslant r \leqslant 0$, put $f(r) = (f(-r))^{-1}$. Show that if r, s and r+s are all in the interval $[-1,1]$ we have $f(r)f(s) = f(r+s)$ and that the mapping $r \to f(r)$ of $[-1,1]$ into G is continuous. Deduce that the mapping can be extended to a non-trivial continuous homomorphism of R into G .

(viii) Hence show that every non-discrete locally compact NSS-group contains a subgroup topologically isomorphic to either R or T .

$$* \ * \ * \ * \ * \ * \ * \ * \ *$$

9 · Non-abelian groups

In this chapter we make a few remarks about non-abelian locally compact Hausdorff groups.

For compact Hausdorff (not necessarily abelian) groups there is a duality theory due to M.G. Krein and T. Tannaka. The dual object of a compact Hausdorff group G is not another topological group, as in the abelian case, but rather the class of continuous finite-dimensional unitary representations of G (or a Krein algebra). For full details, see E. Hewitt and K.A. Ross, Abstract harmonic analysis Vol.II.

Let H be a complex vector space and $T(H)$ the group of all one-one linear transformations of H onto itself. A *representation* of a group G is a map $x \rightarrow T_x$ of G into $T(H)$ such that $T_x \cdot T_y = T_{xy}$, for each x and y in G, with $T_e = I =$ the identity operator. A representation U of a topological group is said to be a *continuous irreducible unitary representation* if (a) H is a Hilbert space, (b) every transformation U_x, $x \in G$ is unitary on H, (c) for every ξ and η in H, the function $x \rightarrow (U_x\xi, \eta)$ of G into the complex numbers is continuous, and (d) there are no proper closed subspaces of H carried into themselves by every U_x, $x \in G$.

The central theorem in representation theory of topological groups is due to I.M. Gelfand and D.A. Raikov.

Theorem (Gelfand-Raikov). *Every locally compact Hausdorff group G has enough continuous irreducible unitary representations to separate points.*

Two special cases are important:

(1) If G is compact then every continuous irreducible unitary representation of G is finite-dimensional (that is, H is a finite dimensional vector space).

(2) If G is abelian then every continuous irreducible unitary representation of G is one-dimensional.

An n-dimensional continuous unitary representation can be thought of as a continuous homomorphism of G into the unitary group $U(n)$. As $U(1) = T$, the one-dimensional representations are just the characters of G . So as corollaries we have the important results of F. Peter, H. Weyl and E.R. van Kampen which were mentioned in Chapters 5 and 6.

Corollary 1. *Let* G *be a compact Hausdorff group. Then for each* $g \in G$, $g \neq e$, *there is a continuous homomorphism* ϕ *of* G *into the unitary group* $U(n)$, *for some* n , *such that* $\phi(g) \neq e$. *Hence* G *is topologically isomorphic to a subgroup of a product of unitary groups.*

Corollary 2. *Every LCA-group has enough characters to separate points.*

The class of locally compact groups which have enough finite-dimensional continuous unitary representations to separate points might be expected to have a pleasant structure theory. They do! Such groups are called *maximally almost periodic*.

Theorem. *Let* G *be a connected locally compact maximally almost periodic group. Then* G *is topologically isomorphic to* $R^n \times K$, *where* K *is compact and connected, and* $n \geqslant 0$.

For information on maximally almost periodic groups, see S. Grosser and M. Moskowitz, Compactness conditions in topological groups, *J. Reine Angew. Math.* 246 (1971) 1-40.

To deal with non-compact non-abelian locally compact
Hausdorff groups we have to turn to Lie groups. For any Lie
group we can define a corresponding Lie algebra and two Lie
groups having the same Lie algebra are locally isomorphic.
(There are a large number of books on Lie groups and Lie
algebras, some are listed in our references.) Further,
locally compact groups can be "approximated" by Lie groups
as follows:

Theorem. *Let* G *be any locally compact Hausdorff group.
Then* G *has an open subgroup* G_1 *such that each open
neighbourhood of* e *in* G_1 *contains a compact normal subgroup* H *such that* G_1/H *is a Lie group.*

Corollary. *Every connected locally compact Hausdorff group
is topologically isomorphic to a subgroup of a product of
Lie groups.*

Finally we mention a structure theorem of K. Iwasawa, On
some types of topological groups, *Ann. of Math.* (2) 50 (1949)
507-557. (The result is derived from the analogous result
for Lie groups using the above approximation theorem.)

Theorem. *Let* G *be a connected locally compact Hausdorff
group. Then* G *has a maximal compact subgroup, and all such
subgroups are connected and conjugate to each other. Let* K
be one of them. Then G *has subgroups* $H_1, \ldots, H_n$ *each
topologically isomorphic to* R *and such that each element
$g \in G$ can be decomposed uniquely and continuously in the
form* $g = h_1 \ldots h_n k$ *, with* $h_i \in H_i$ *and* $k \in K$ *. In
particular,* G *is homeomorphic to* $R^n \times K$ *.*

In the abelian case this theorem reduces to Theorem 26.

References

1. N. Bourbaki, *General Topology*, Parts I and II, Addison-Wesley, 1967.

2. J. Dieudonne, *Treatise on Analysis*, Vol.II, Academic Press, 1970.

3. E. Hewitt and K.A. Ross, *Abstract Harmonic Analysis*, Vols.I and II, Springer-Verlag, 1963.

4. P.J. Higgins, An Introduction to Topological Groups, *London Math. Soc.* Lecture Note Series No.15, 1974.

5. T. Husain, *Introduction to Topological Groups*, Saunders Co., 1966.

6. I. Kaplansky, *Lie Algebras and Locally Compact Groups*, Chicago Univ. Press, 1971.

7. L.H. Loomis, *An Introduction to Abstract Harmonic Analysis*, Van Nostrand, 1953.

8. G. McCarty, *Topology - An Introduction with Applications to Topological Groups*, McGraw-Hill, 1967.

9. D. Montgomery and L. Zippin, Topological Transformation Groups, *Interscience*, 1955.

10. M.A. Naimark, *Normed Rings*, Noordhoff, 1959.

11. L.S. Pontryagin, *Topological Groups*, Gordon and Breach, 1966.

12. W. Rudin, Fourier Analysis on Groups, *Interscience*, 1967.

Lie Groups

13. C. Chevalley, *Theory of Lie Groups*, Princeton Univ. Press, 1946.

14. P.M. Cohn, *Lie Groups*, Cambridge Univ. Press, 1967.

15. S. Helgason, *Differential Geometry and Symmetric Spaces*,

Academic Press, 1962.

16. G. Hochschild, *The Structure of Lie Groups*, Holden-Day, 1965.

17. J. Price, Lie Groups and Compact Groups, *London Math. Soc.* Lecture Note Series No.25.

18. A.A. Sagle and R.E. Walde, *Introduction to Lie Groups and Lie Algebras*, Academic Press, 1973.

19. J.-P. Serre, *Lie Algebras and Lie Groups*, Benjamin, 1965.

Index of terms

annihilator 90ff

approximation by Lie groups 118

approximation by metrizable groups 63

arcwise connected 16

Ascoli's theorem 45

Baire category theorem 22

Banach space 14,48

basis 33

Bohr compactification 88,101

box topology 17,69

$C(-, -)$ 43

cardinality of character groups 96,102f

cartesian product 1,16

character(s) 47

   extendability 66,91

   enough 59-62,75f,70,87,116f

character group 47

circle group 1

closed subgroups of R^n 33

   T^n 36

   $R^n \times Z^m \times K$ 36,94

commutator subgroup 15,22

compact element 97-99

compact group, dual is discrete 50

compact open topology 42

compact torsion groups, characterized 69

for non-locally compact groups 68f

torsion group, compact characterized 69

torsion-free group 21,53,87,99f,103,108,110

torsion subgroup 99

totally disconnected 13

Tychonoff topology 17

Tychonoff theorem 17

uniform space 41

uniformity 41

unitary group 2

unitary matrix 2

unitary representation 116f

weak direct product 17f

Z 1,47,51,55

    character group of 47,51

Index of exercises, propositions and theorems